xLV, 5, 129

D1145539

Computers in Chemistry

Pete Biggs

Computer Manager, Physical & Theoretical Chemistry Laboratory, Oxford

Series sponsor: ZENECA

ZENECA is a major international company active in four main areas of business: Pharmaceuticals, Agrochemicals and Seeds, Specialty Chemicals, and Biological Products.

ZENECA's skill and innovative ideas in organic chemistry and bioscience create products and services which improve the world's health, nutrition, environment, and quality of life.

ZENECA is committed to the support of education in chemistry and chemical engineering.

OXFORD
UNIVERSITY PRESS

OXFORD
UNIVERSITY PRESS

Great Clarendon Street, Oxford OX2 6DP
Oxford University Press is a department of the University of Oxford.
It furthers the University's objective of excellence in research, scholarship,
and education by publishing worldwide in

Oxford New York

Athens Auckland Bangkok Bogotá Buenos Aires Calcutta
Cape Town Chennai Dar es Salaam Delhi Florence Hong Kong Istanbul
Karachi Kuala Lumpur Madrid Melbourne Mexico City Mumbai
Nairobi Paris São Paolo Singapore Taipei Tokyo Toronto Warsaw

with associated companies in Berlin Ibadan

Oxford is a registered trade mark of Oxford University Press
in the UK and in certain other countries

Published in the United States
by Oxford University Press Inc., New York

© Pete Biggs, 1999

The moral rights of the author have been asserted
Database right Oxford University Press (maker)
First published 1999

All rights reserved. No part of this publication may be reproduced,
stored in a retrieval system, or transmitted, in any form or by any means,
without the prior permission in writing of Oxford University Press,
or as expressly permitted by law, or under terms agreed with the appropriate
reprographics rights organization. Enquiries concerning reproduction
outside the scope of the above should be sent to the Rights Department,
Oxford University Press, at the address above.

You must not circulate this book in any other binding or cover
and you must impose this same condition on any acquiror

A catalogue record for this book is available from the British Library

Library of Congress Cataloging in Publication Data
(Data available)

ISBN 0 19 850446 2

Typeset by the author

Printed in Great Britain
on acid-free paper by
Bath Press, Avon

Series Editor's Foreword

Oxford Chemistry Primers are designed to provide clear and concise introductions to a wide range of topics that may be encountered by chemistry students as they progress from the freshman stage through to graduation. The Physical Chemistry series contains books easily recognised as relating to established fundamental core material that all chemists need to know, as well as books reflecting new directions and research trends in the subject, thereby anticipating (and perhaps encouraging) the evolution of modern undergraduate courses.

In this Physical Chemistry Primer Peter Biggs has produced an authoritative and easy-to-read introductory account of Computers in Chemistry. The Primer presents essential basic ideas required to understand and exploit computers as encountered by chemistry students in their studies and in the laboratory at all stages up to and including research level. This Primer will be of lasting and broad value to all students of chemistry (and their mentors).

Richard G. Compton
Physical Chemistry Laboratory,
University of Oxford

Preface

A chemist's life would be very difficult without computers, and these days it would be difficult to find a chemist without at least one computer in their laboratory: computers have become an integral part of chemistry. This 'invasion' is not constrained to the beige box on the desk. Virtually all modern instrumentation contains some form of computer, and indeed the operation of many instruments has become so complex that it is impossible without some form of computer control. So, it is important that a modern chemist has at least some knowledge of computers, and the deeper that knowledge is, the better use will be made of the tools available.

This Primer provides an overview of computers and their use in chemistry. It is not a programming manual, nor is it meant as a definitive guide to interfacing or document processing – there is just not enough space here for that. What I do hope though is that it will help the reader to gain a better insight into the workings of a computer and so help them use the facilities available to them more effectively.

Thanks must go to Prof. Richard Wayne, who gave me the opportunity to pursue my love of electronics and computers whilst still calling myself a chemist. But, primarily, I must thank my wife, May, for suffering most during the production of this book, and for producing our beautiful daughter, Pippa.

Oxford   P. B.
1999

Contents

1 Introduction

Computers and chemistry have been linked ever since useful calculating machines were invented. Today, you are hard-pressed to find a research room that doesn't contain a computer of some sort; indeed whole branches of chemistry have been made possible through the use of digital computers.

The use of computers in chemistry is not restricted to computation though: they are used in applications ranging from control of equipment and data acquisition to accessing databases both locally and on remote systems. This primer gives the reader an introduction to these many varied aspects of the use of computers in chemistry.

1.1 A brief history of computers

In order to fully understand some of the concepts introduced later, it is helpful to have a basic knowledge of the history of computing.

The original meaning, and one that is still valid, of 'computer' is "one who computes". Indeed the Oxford English Dictionary still lists this definition before all others. The contemporary usage of the word 'computer' is not actually that modern: the first recorded use of the word to mean a machine to perform computations was in the late nineteenth century. However, it was not until midway through the twentieth century that it came to mean an electronic machine capable of being programmed to perform a predefined sequence of calculations.

Computer: 1. One who computes; a calculator, reckoner; *spec.* a person employed to make calculations in an observatory, in surveying, etc.
2. A calculating-machine; esp. an automatic electronic device for performing mathematical or logical operations.
(source: Oxford English Dictionary, New Edition)

The father of mechanical calculating machines must be Charles Babbage. He built a prototype of his Difference Engine in 1822 and shortly after, in 1834, he conceived the first automatic digital computer, the Analytical Engine. Babbage's machine contained many elements that became building blocks of modern computers, such as an arithmetic unit, a memory for storing numbers, an input/output medium, and sequential control. Unfortunately Babbage never completed the Analytical Engine, largely due to lack of funds and insufficient precision in the machining methods of the day; the device was eventually constructed successfully in 1991 and now resides in the Science Museum, London.

Babbage was not the first to invent a machine for performing calculations. As early as 1623, Wilhelm Schickard, a friend of the astronomer Kepler, invented the first mechanical calculator, although the records of this machine were subsequently lost in the Thiry Years' War. Later, in 1642, Blaise Pascal invented a machine that could perform addition and subtraction, and in the 1670s the German mathematician Liebniz developed a more advanced calculator that was able to multiply, divide, and calculate square roots.

The next major step in computer development was that of the Frenchman Joseph-Marie Jacquard. In 1804, he developed an automatic loom in which the woven pattern was controlled by a series of punched cards. Babbage was later to use the idea of punched cards in his Analytical Engine.

After Babbage, the next significant milestone was the work of the English logician George Boole. One of Boole's theories, first published in 1847, concerned the application of logical operators (e.g. AND, OR, and NOT) to binary numbers. This theory formed the elements of *Boolean Algebra*, the basis on which the circuits within modern digital computers operate.

The American census of 1890 provided another, unlikely, milestone. Herman Hollerith, an American statistician working for the U.S. Postal Service during the 1880 census, conceived of an electromechanical machine that could sort and tabulate the census data; using the machine resulted in the data being processed in one-third the time normally required. Hollerith founded the Tabulating Machine Company in 1896; this company later became the International Business Machines Corporation, more commonly known as IBM.

1.2 Computers in chemistry

There has been a long and fruitful association between chemistry and computing. To a large extent it has been a symbiotic relationship: not only has the presence of computers opened-up new areas of chemical research, but the needs of chemists has been a major contributory factor in the development of larger and faster computers.

The first large users of computer power within the chemistry community were the crystallographers. In the 1950s, developments in crystallography had slowed because of an inability to rapidly and reliably perform the calculations necessary to derive a structure. The development of computers meant that the required calculations could be completed and that particular branch of science was revitalised.

Chemistry and computing became so entwined that in many universities it was the chemistry department that was the impetus behind the installation of sizeable computer facilities. Indeed, even to this day in Oxford, computing is filed under chemistry in the University administration, since it was under the auspices of the chemistry department that the first computer facility was installed.

It wasn't long before crystallographers weren't the only chemists interested in computers. Many traditional fields of physical and inorganic chemistry, such as spectroscopy, were revitalised by computers, and yet others, such as theoretical chemistry or molecular modelling, owe their whole existence to computers. Computers have not just had an impact because of their computing power, many modern experiments, in which data is collected by computers, would have been very different in the pre-computer age.

Over the last 10 years or so, computers have had an even more profound effect on the life of a chemist. Not only are computers used as part of the experimental procedure, but they are now also vital to all aspects of chemistry: from researching the background to a piece of work using the Internet, to producing and publishing the final results, from passing email between co-workers to virtual conferences.

1.3 Introduction to digital electronics

It would be impractical to try and give a full introduction to electronics here, but in order to fully understand some of the material in later sections it is necessary that the reader has at least some knowledge of electronics. Much of the groundwork is covered in Wayne's primer *Chemical Instrumentation*, with a more advanced coverage being provided by Horowitz & Hill. It will be assumed that the reader of this primer has a basic knowledge of the 'physics' of electronics, *i.e.* resistance, capacitance, and inductance; further, it is assumed that a basic knowledge of semiconductors such as diodes and transistors is held.

The area of electronics concentrated on here is that of *digital electronics*. Digital electronics is, to a large extent, much simpler than *analogue* electronics – it is usually just a matter of connecting basic building blocks together to produce the desired result. Digital electronics is also much simplified by the fact there are only two possible states: either 'on' or 'off'. These two states are often called '1' and '0' or 'high' and 'low' or 'true' and 'false'.

Logic gates

The basic units of digital electronics are called *gates*; these are devices that have one or more inputs and a single output, the state of the output is determined by the combination of states of the inputs. The basic gates are: buffers, OR, and AND; their negative counterparts are: inverters (NOT), NOR and NAND. The basic operation of these gates follows Boolean logic (indeed they are often called *logic gates*), *i.e.* the output of a two input AND gate is on only if input A **and** input B are both on, similarly the output of a two input OR gate is on if either input A **or** input B is on. The NAND and NOR gates are similar except that the output is inverted, *i.e.* the output of a NAND gate is off only if input A **and** input B are both on.

The operation of a logic gate, or a group of gates, is generally described by a *truth table*. The truth table lists all possible combination of inputs and the corresponding outputs. The truth tables, along with the associated symbol and logical function, for each of the basic gate types is shown in Fig. 1.1; a ring (o) on an input or output of a symbol shows inversion.

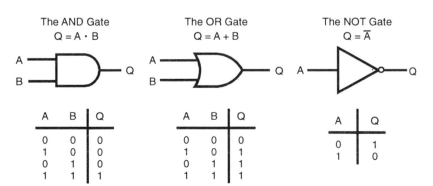

Fig. 1.1 The three basic logic gates, their symbols, and associated truth tables

More complex gates can be built out of these basic gates. Take for example the logical exclusive-OR (or XOR) gate. With the XOR function the

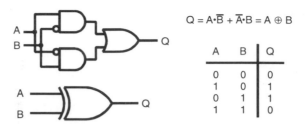

$$Q = A\cdot\overline{B} + \overline{A}\cdot B = A \oplus B$$

A	B	Q
0	0	0
1	0	1
0	1	1
1	1	0

Fig. 1.2 The XOR gate

output is high if, and only if, only one of the two inputs is high. The logical expression of this functionality is thus $\overline{A}B + A\overline{B}$ showing immediately how the XOR gate can be made from a combination of NOT, AND and OR functions (Fig. 1.2). The XOR function is widely used and so is given its own mathematical ($\oplus$) and logical symbol (Fig. 1.2). In actual fact though, nearly all logic operations can be performed using solely NAND and NOT gates: this can be illustrated by the construction of OR gate simply by inverting the inputs to a NAND gate (Fig. 1.3).

Fig. 1.3 The OR function constructed using the NAND gate

$$Q = \overline{\overline{A}\cdot\overline{B}} = A \times B$$

A	B	$\overline{A}$	$\overline{B}$	$\overline{A}\cdot\overline{B}$	Q
0	0	1	1	1	0
1	0	0	1	0	1
0	1	1	0	0	1
1	1	0	0	0	1

The term 'TTL' refers to the type of electronics and construction used in the design of the integrated circuit and stands for Transistor-Transistor Logic. Other types are DTL (Diode-Transistor Logic) and CMOS (Complementary Metal Oxide Semiconductor).

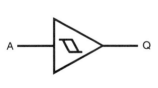

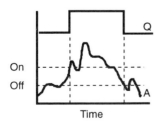

Fig. 1.4 The Schmitt trigger

Although these digital circuits generally operate at two discrete levels (*i.e.* on or off), they are made up of analogue components. In fact the gates can often be thought of as high-gain amplifiers. One consequence of this is that input voltage levels between the nominal 'on' or 'off' states can cause unpredictable behaviour of the output. Take for example TTL logic gates: the 'off' level is defined to be <0.8V, whereas the 'on' level is >2V. The gap between on and off levels is not an issue when gates are connected together, but may be a problem when interfacing to the outside world. The solution to this is to use a device called a *Schmitt Trigger*. A Schmitt Trigger is a logical element that switches state at predetermined levels; for instance in TTL Schmitt devices, the positive-going threshold when the device turns on is 1.7V, whereas the negative-going threshold, when the device turns off, is 0.9V. This *hysteresis* of 0.8V is important in interfacing noisy, slowly changing signals to digital circuits (see Fig. 1.4).

Bistable devices

In the devices described so far the output state follows exactly the conditions of the inputs, *i.e.* if the output goes high because an input goes high, then when that input goes low again, the output will go low. However, there is a group of devices in which this is not always the case – the so-called bistable devices. These devices, as their name suggests, are stable in two distinct states; that is to say, the output state is determined by a special combination of the inputs, and when the input conditions are changed, the output will remain in that state. The output state can then be changed back by a different set of input conditions. These bistables are often called *flip-flops* as they can be flipped and flopped from one state to the other.

The simplest bistable is the R-S flip-flop. The name 'R-S' comes from the fact that these devices can be *set* or *reset* to a particular state. From the truth table shown in Fig 1.5, it is evident that when either input R or S goes high, the output will go high or low respectively; however when both R and S are low, the output will remain in whatever state it was in before the inputs went low. Typically R-S flip-flops are used for switch debouncing: when any mechanical switch is activated there is a very short period of time when the contacts bounce after they are closed (the effect may be that the switch appears to open and close rapidly for a millisecond or so), this bounce can cause many problems with logic circuits especially those that 'count' how many times a switch is depressed. The debouncing is achieved by connecting each input of the R-S flip-flop to one pole of the switch and arranging for these inputs to also be connected to ground *via* a pull-down resistor (see Fig. 1.6). Now, when the switch is moved from the R to the S position, the first contact of the switch will cause the flip-flop to change state; any bounces of the switch will only cause input S to go low, it will not cause input R to go high, thus the output will remain steady – at least until the switch is returned to the R position.

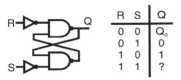

R	S	Q
0	0	Q_0
0	1	0
1	0	1
1	1	?

Fig. 1.5 The R-S flip flop. Q_0 is the value of Q before the inputs attained their present state.

The next level of sophistication of the flip-flop is the D-type. These flip-flops are basically data latches (hence the name 'D-type'). A latch is a device whose outputs can be locked into a state that reflects the inputs on a given signal. Hence they are used to effectively 'freeze' a signal at a particular time, and are very crude memory devices. In the device shown in Fig 1.7, the output remains in the same state so long as the CLK input is low; however when the CLK input goes high, the Q output is set to the same state as the D input. In this particular example, the Q output will continue to mimic the D input for as long as the CLK input is high. In other similar devices, the information is transferred from the D input to the Q output only when the CLK input performs a low to high transition; such devices are said to be *edge-triggered*. The advantage of edge-triggering a device is that events can be synchronised within a circuit much more easily than with level triggering (hence the labelling of such inputs as 'clocks').

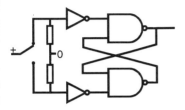

Fig. 1.6 The R-S flip-flop as a switch debouncer

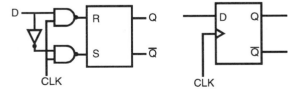

Fig. 1.7 The D-type flip-flop

The final type of flip-flop we will consider here is the J-K flip-flop. Like the others described above, data can be latched into the device, but in this case, if both the inputs are high, the output *toggles* on a clock edge, i.e. it changes to the opposite state (Fig. 1.8). If the clock input is regularly

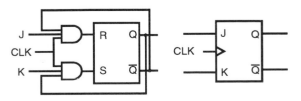

Fig. 1.8 The J-K flip-flop

changed, the output will change at half the frequency of the clock. Further, if the output of the flip-flop is connected to the clock input of another, the output of the second one will change at a quarter of the frequency of the original clock. We have, therefore, the basic device for counting or dividing. The flip-flops can be *cascaded* to whatever depth we require to provide either a 2^n divider or a counter with a maximum count of 2^n (Fig. 1.9).

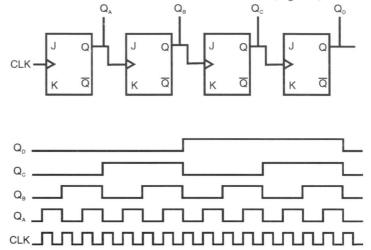

Fig. 1.9 Cascaded J-K flip-flops. This arrangement is the basis of a counter. The lower diagram shows the state of each output at each successive clock pulse.

It should be stressed that the schematics of the flip-flops shown here are simplified to illustrate their basic properties, there are many refinements used in real devices to minimise things such as clock edge effects and propagation delays, and to provide extra features such as preset and clear functions.

One further use for flip-flops, which is of importance in computing, is in the construction of *shift registers*. Shift registers find many uses in computing and digital electronics: a binary number shifted right or left is mathematically equivalent to multiplication or division by 2 respectively; the action of shifting a group of bits is the basis of converting a parallel signal to a serial signal (and *vice versa*); and so on. The basic shift register is made up of a series clocked R-S flip-flops, the Q and $\overline{Q}$ outputs of one connected to the S and R inputs of the next; the clocks are all connected together (see Fig. 1.10). On one of the clock edges (typically the positive going edge), the

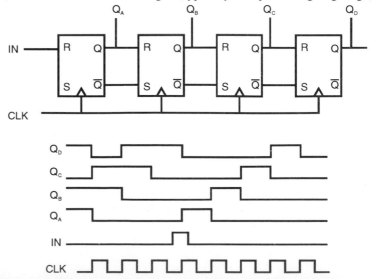

Fig. 1.10 RS flip-flops connected as a shift register. Each output takes on the value of the previous one on each positive going edge of the clock. The effect is to shift the data along one position on each clock pulse as can be seen by the single pulse on the input *rippling* along the outputs.

information on the inputs of each flip-flop is transferred to the outputs, and the information is shifted along one place: output Q_D takes on the value of Q_C, Q_C takes on the value of Q_B and so on, Q_A takes the value present at the serial input. The serial output at Q_D will be the same as the serial input, only delayed by 4 clock cycles, hence creating a *delay line*. The outputs Q_A to Q_D will reflect the state of the serial input during the last 4 clock cycles (Q_A being the most recent), resulting in a serial to parallel converter, similarly the output at Q_D will be a serial representation of Q_A to Q_D.

1.4 Number systems

As we have already seen, computers and logic circuits operate as binary devices, *i.e.* their outputs can be in one of two states. Consequently, in order that more than two discrete states may be represented at any one time, a number of these single binary devices can be grouped together. Most often these groups of binary outputs are interpreted as representing numbers and are called *binary numbers*; each individual digit of these numbers are called *binary digits*, or *bits*.

Binary numbers work in a similar way to the more familiar decimal numbers. In school we learnt that the number 294 means {2 hundreds + 9 tens + 4 units}, with each digit representing a power of 10 higher than its right hand neighbour.

$$2 \times 10^2 + 9 \times 10^1 + 4 \times 10^0 = 294$$

The only thing different about binary numbers is that each successive place to the left represents a power of 2 higher, and the only digits possible are 0 and 1. Thus the binary number 11011 means in decimal

$$
\begin{aligned}
&1 \times 2^4 &&+ 1 \times 2^3 &&+ 0 \times 2^2 &&+ 1 \times 2^1 &&+ 1 \times 2^0 \\
={}&16 &&+ 8 &&+ 0 &&+ 2 &&+ 1 \\
={}&27
\end{aligned}
$$

There is, obviously, scope for much confusion between, for example, the binary number 10 and the decimal number 10. Where there is possible ambiguity, it is usual to distinguish between them by adding a letter 'B' after the binary number and a letter 'D' after the decimal number. Thus 10B=2D and 11011B=27D.

As binary numbers get larger, the number of digits required to represent them rapidly gets too large to handle. A 16-bit binary number can only represent up to 65535 in decimal, but is already very cumbersome to read and write. Consequently it is usual to split the binary number into groups of digits and represent those groups as numbers. The usual grouping is either three or four bits, resulting in groups representing either a maximum of 8 or 16 in decimal. The most convenient method of writing these groups is as a single digit, resulting in either base-8 or base-16 numbers, or, as they are more commonly called, *octal* or *hexadecimal*. Octal numbers are not used very often now, so we will concentrate on hexadecimal numbers.

Hexadecimal

A single hexadecimal digit can represent a decimal number from 0 to 15. It is thus difficult to represent all the possible hexadecimal digits using only

Hexadecimal numbers are indicated by suffixing the number with 'H', *i.e.* 32H. They may also be prefixed by '0x', as in 0x32, or '#', as in #32, or &H as in &H32 depending on the standard practices in particular circumstances.

decimal digits. The solution is to extend the range of characters used to represent the digits by including the first six alphabetic characters, *i.e.* 'A' to 'F'. 'A' thus stands for 10, 'B' for 11 and so on up to 'F', which stands for 15 in decimal. Hexadecimal numbers work in the same way as any other number base, except that each place to the left in the hexadecimal number is a higher power of 16. Thus the hexadecimal number 3FA04 is equivalent to the decimal.

$$3 \times 16^4 + F \times 16^3 + A \times 16^2 + 0 \times 16^1 + 4 \times 16^0$$
$$= 3 \times 16^4 + 15 \times 16^3 + 10 \times 16^2 + 0 \times 16^1 + 4 \times 16^0$$
$$= 196608 + 61440 + 2560 + 0 + 4$$
$$= 260612$$

Initially the concept of using letters in numbers may be a bit strange, but with use, it will become natural. Converting between binary and hexadecimal is very easy. Since each hexadecimal digit represents four binary digits, all that is required is to concatenate the binary representation of each hexadecimal digit in order to construct the binary number:

3	F	A	0	4
0011	1111	1010	0000	0100

So the binary number equivalent of 3FA04 is 00111111101000000100. Going the other way is just as easy: the binary number is split into groups of four bits and the hexadecimal number equivalent of each group is written down.

Negative numbers

The representation and manipulation of negative numbers in binary and hexadecimal is often a cause of problems. In decimal, a sign indicator (*i.e.* the minus sign '−') is placed in front of a number to indicate that its value is less than zero. This is fine for human readers, but it makes mathematics difficult as it is full of special cases and rules to help deal with the sign indicator. In binary, life is easier as there are usually only a fixed number of bits available. Thus, in decimal, when you reduce the value of 0 by 1 you get −1, whereas in binary, if you reduce 0000 by 1 you get 1111, *i.e.* the number *wraps round*. This may at first seem odd, translating these numbers into decimal effectively means that reducing 0 by 1 gives 15!

The situation can be made much clearer by considering that in a *signed* binary number, the most significant digit of the binary number represents its negative decimal equivalent. So, in our 4-bit number, the most significant bit represents not 2^3, but -2^3. So, 1111 represents not 15, but

$$1 \times (-2^3) + 1 \times 2^2 + 1 \times 2^1 + 1 \times 2^0$$
$$= -8 + 4 + 2 + 1$$
$$= -1$$

It is thus very important when dealing with binary numbers to know if they represent *signed* or *unsigned* numbers. This method of representing numbers in binary is called *two's complement* and is the most common method.

It is relatively easy to calculate the two's complement of any number, all that is necessary is to invert each of the digits in the positive equivalent of the number, then add one (forgetting any carry digit). Thus, the 8-bit two's complement representation of –37 is given by taking the binary equivalent of +37, 00100101, inverting all the digits to give 11011010, then adding 1 to give 11011011. In order to verify that this is indeed the correct representation of –37, we can add 37 to it and we should get zero, since –37 + 37 = 0. If we perform this binary addition (*i.e.* 11011011 + 00100101) we get as the result 100000000, which, as this is an 8-bit system, is zero.

Table 1.1 shows the 8-bit binary and hexadecimal equivalents of some positive and negative numbers. From this it is obvious that a negative number always has the most significant bit set, but it must be stressed that this is not a sign indicator in the same way as the negative sign is in decimal numbers: it is a useful consequence of the number system used.

It was mentioned briefly previously that it is important to know if a binary number is signed or not. There are a few situations where this becomes particularly critical. As we will see in later chapters, the data collected from digital devices is, for example, 8-bit. This would mean that a voltage, say, is represented as having 256 discrete steps ranging from 00H to FFH. If this number were stored in a variable in a program that is interpreted as a signed number, then the program will treat this number incorrectly. A voltage represented by 7FH would be correctly interpreted as 127D, but just a small amount higher voltage will be represented by 80H, which would be incorrectly interpreted as –128D. Any mathematics performed using such incorrectly interpreted numbers will, obviously, be wrong.

This *wrapping* of signed numbers from a large positive value to a large negative value is not just an inconvenience. By the time that many will be reading this book, the *Millennium bug* or *Y2K bug* will have been and gone. (Indeed, if the prophets of doom are correct, there may be no one reading this book!) But there is another less publicised date problem in the future. Many Unix systems represent the date as a number of seconds since January 1st 1970; this number of seconds is most often held as a 16-bit number. Unfortunately, that number is a signed number. Sometime in early 2038, that signed number will reach 7FFFFFFFH, or about 2.15×10^9 seconds. The next second, the number will be 80000000H, or, because it is a signed binary number, -2.15×10^9. At that point the date will be interpreted as sometime in 1902 with obvious consequences. The date in 1902 that this occurs on happens to be a Friday 13th!

Fractional numbers

So far we have only dealt with whole numbers, or *integers*. In most scientific work it is necessary to use some form of fractional, or *real* numbers. In decimal, fractional numbers are represented as powers of 10 less than zero. So the number 178.125 is actually $1 \times 10^2 + 7 \times 10^1 + 8 \times 10^0 + 1 \times 10^{-1} + 2 \times 10^{-2} + 5 \times 10^{-3}$. Alternatively the so-called *scientific* format may be used in which the number is represented as a *mantissa*, whose value is between 1 and 10, and *exponent* signifying the power of 10 that the mantissa needs to be multiplied by to obtain the original number. So, the number 178.125 can be written as 1.78125×10^2, or 1.78125E2.

All these formats are equally as acceptable for binary numbers. For instance the decimal number 178.125 can be represented as $10110010_\Delta 001$, that is

Table 1.1 The binary and hexadecimal equivalent of some signed numbers		
+127	01111111	7F
+126	01111110	7E
.		
+37	00100101	25
.		
+2	00000010	02
+1	00000001	01
0	00000000	00
–1	11111111	FF
–2	11111110	FE
.		
–37	11011011	DB
.		
–127	10000001	81
–128	10000000	80

In the UK it is thought that a Friday that is the 13th in the month is an unlucky day.

The symbol '$_\Delta$' is used to show the transition point between units and fractions in binary numbers, the same as '.' is used in decimal numbers.

$1 \times 2^7 + 0 \times 2^6 + 1 \times 2^5 + 1 \times 2^4 + 0 \times 2^3 + 0 \times 2^2 + 1 \times 2^1 + 0 \times 2^0 + 0 \times 2^{-1} + 0 \times 2^{-2} + 1 \times 2^{-3}$

$= 128 + 32 + 16 + 2 + \frac{1}{8}$

$= 178.125$

However these straight binary fractions are not really suitable for use in calculations since in order to represent a number of any significant size would require many bits. So, in the same way as scientific notation can be used to represent decimal numbers, binary scientific notation can be used to represent binary numbers. Taking the same binary number as before, this can be represented as $1_\Delta 0110010001 \times 2^{111}$ (where all the numbers are binary), or $1_\Delta 0110010001E111$.

One complication is the matter of negative numbers, both in the mantissa and the exponent. The use of two's complement for the mantissa is of no benefit in this case, since any mathematics can not be performed directly because of the presence of the exponent. Consequently the representation of the number in the computer usually just includes a *sign-bit*. This is a single bit, usually the most significant bit, that when set indicates that the mantissa is negative.

Negative exponents, *i.e.* those representing numbers less than one, are often expressed as a *biased* or *offset* two's complement number. This just means that the exponent is expressed as a two's complement number, then a fixed offset is added to it. The constant added to the exponent depends on the total number of bits allocated for its storage, if this is 8 bits, then constant is 127 (7FH). We can thus see that an exponent of 111B, as above, will be represented as 10000110. The reasoning behind using such a scheme is that it makes comparing magnitudes much easier than if a straight two's complement system is used.

The most common format for storing real numbers in computers is the *IEEE format*. This takes all the above principles and packs them in a well-defined format into a binary number. There are two sizes of IEEE number: short real and long real. The short real uses a 23-bit mantissa, an 8-bit exponent and a sign bit, making a total of 32 bits. The long real on the other hand is a 64-bit number consisting of a sign bit, 52-bit mantissa and 11-bit exponent. In both these formats the mantissa is expressed without the leading 1, since all mantissas are between 1 and 2 (*i.e.* 1B and 10B) and so they all start with 1_Δ. The range of decimal numbers representable by the short and long integers and the number of significant decimal digits is shown in Table 1.2.

Table 1.2 Floating point data types

Data Type	Bits	Significant decimal digits	Approximate decimal range
Short Real (single precision)	32	6–7	$8.43 \times 10^{-37} \leq |x| \leq 3.37 \times 10^{38}$
Long real (double precision)	64	15–16	$4.19 \times 10^{-307} \leq |x| \leq 1.67 \times 10^{308}$

It should also be remembered that just as there are decimal fractions, such as $\frac{1}{3}$, that can not be represented exactly, there are binary fractions that also can not be represented exactly. One such is '0.1': the 16-bit binary fraction representing 0.1 is $1_\Delta 1001100110011001E100$, but this number only approximately equals 0.1, it being about 5.7×10^{-7} less. Such rounding errors may become significant when, for instance, performing subtractions on two very close numbers, but the large number of bits in the IEEE format numbers means that the difference between such irrational binary numbers and their decimal equivalent is kept very small.

The final problem with manipulating real numbers is that of exceptional values. Table 1.2 shows the range of values that the real numbers can take: notice that zero is not included in any of the ranges. This is because zero is represented absolutely by a mantissa and exponent of all zeroes. Surprisingly, it is possible to have a positive zero or a negative zero; the sign of zero is significant in some operations. Other special numbers are $+\infty$, $-\infty$ and *indefinite*. The $\pm\infty$ is generated when, for example, division by zero is attempted; *indefinite* is a result of operations such as $\sqrt{-1}$. Each of these conditions is represented by specific combinations of mantissa and exponent. Often, when dealing with floating point numbers, you may see 'NAN', this means 'not a number' and results from an invalid combination of bits. The individual cases just mentioned are special classes of NANs.

2　The hardware

2.1 What is a computer?

This is actually a very difficult question to answer. A computer is different things to different people. To most people, a computer is a beige box sitting on a desk; a computational chemist may think of it as an anonymous series of cabinets at the other end of a cable; to others it is a black piece of plastic containing silicon. In actual fact, a computer is all of these things, and more. In Chapter 1 some background to the history of computers was presented, and some insight into how computers developed was given. Also in that chapter was a brief introduction to digital electronics: the building blocks of computers. Now we delve into what a computer is physically made up of.

Elements of a computer

The basic elements of a computer haven't changed since Babbage's time, although they have certainly changed in their implementation. Fig. 2.1 shows a very simplified block diagram of a computer. The main elements are a *central processor unit* (*CPU*) that does the work, *memory* to store results and the controlling program, *input* and *output* (*I/O*) devices to communicate with the outside world, and *buses* to provide the communication between the various elements. In the following few sections we will look at each of these in more detail.

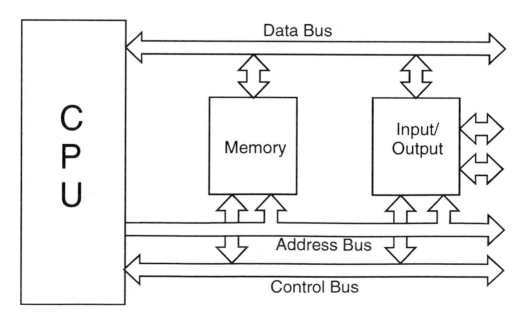

Fig. 2.1 The basic elements of a computer

Memory and buses

It may seem strange to start by describing memory, rather than the CPU, but it is memory that makes a computer unique. Memory can be thought of as a series of locations into which information can be put; with each of these locations having an address by which it is known. The specific memory location required by the CPU is controlled by the *address bus*, and the data which is read or written to that location is on the *data bus*; the type of access (*i.e.* read or write) is controlled by the *control bus*.

There are a number of different types of memory commonly in use. *Read-only memory (ROM)* is memory whose contents are permanent, *i.e.* it can only be read, and not (usually) written to, and the contents do not disappear when power is removed. Typically ROM is used to store permanent programs such as those required when the computer is first switched on. *RAM (random access memory)* is memory that can be both read from and written to. In an application where the computer is a dedicated controller, the amount of RAM need only be very small as it is only used for storing data as it is manipulated by the CPU. However, in most modern computer systems, the vast bulk of memory is RAM; RAM in this case holds both data that is being processed by the CPU and the program code controlling that processing. Unfortunately RAM, by its very nature, is volatile, and the contents of RAM are lost when power is removed from the memory.

The maximum amount of memory in a computer is determined largely by the size of the buses. In early CPUs, the address bus was 16-bits, or less, wide (*i.e.* there were 16 lines making up the bus) meaning that a maximum of 65,535 (64k) locations could be accessed, whilst the data bus was at most 8-bits wide. The processor used in the Acorn BBC Micro (the Motorola 6502) falls into this class. However, the first IBM PCs used either an 8086 or 8088 CPU (both from Intel) which have a 20-bit wide address bus with either a 16- or 8-bit wide data bus respectively. Consequently the 8086 can address up to 1,048,576 (1024k) locations. This address space seemed very large at the time, especially when you consider that memory was probably the largest cost single item in a computer, but it is dwarfed by the massive address spaces available in modern computers. The most recent Intel processor (the PII) has a 32-bit address bus giving an address space of 4Gb, with many of the CPUs used in larger machines capable of addressing over 1Tb of memory; no doubt in the future, these memory sizes will seem small.

There are other, more specialised, types of memory commonly found in computers that should be mentioned here. One of the features of memory is that for most of the time, it is quiescent: the CPU can usually only address one memory location at a time, and in general it will access the memory sequentially. So, rather than use expensive high speed memory for all the memory in the computer, lower cost slower memory is mostly used, with a small amount of high speed memory used to act as temporary store for the data that the CPU is currently accessing. This temporary memory is called *cache*. Since it is accessed at very high speeds, cache is usually located very close to the CPU, both logically and physically, indeed most CPUs now come with cache located on the same chip as the CPU. The control circuitry for the cache normally loads large sections from main memory into cache each time that a new memory area is accessed; the principle behind this behaviour is that on average, when the CPU accesses a memory location, it

The unit multipliers used in computing are slightly different from those specified for SI units! The basic unit is a '*bit*' (standing for *binary digit*), 8 bits is called a '*byte*', with 4 bits often being called a '*nibble*'. The abbreviation for bit is 'b', and that for byte is 'B'.

$$2^{10} = 1024 = 1k$$
$$2^{20} = 1024 \text{ k} = 1M \text{ ('mega')}$$
$$2^{30} = 1024 \text{ M} = 1 \text{ G ('giga')}$$
$$2^{40} = 1024 \text{ G} = 1 \text{ T ('tera')}$$

Hence 2MB is 2048kB or 2,097,152 bytes.

will access the next location in sequence. When the cache is full, the oldest unused data is written back to main memory and the area re-used. The detailed operation of cache is beyond the scope of this book, but it will become evident in later chapters that the amount and organisation of any cache is of critical importance in designing efficient programs.

There are also types of bus other than the basic ones mentioned so far. Of most importance to us here is the fact that some processors have a separate bus dedicated to I/O operations. This I/O bus is generally slower than the main buses (since I/O devices are generally slower than the CPU), and is not as large. For instance, the 8086 series of Intel CPUs have a 16-bit wide I/O address bus allowing up to 64k locations to be addressed (there is no separate I/O data bus). The I/O bus is used to communicate with devices attached to the outside world, such as video, serial lines, printers and, very importantly for our purposes, experiments.

The central processor unit

The central processor unit, as its name suggests, is the key element of a computer. The CPU contains an *arithmetic-logical unit* (*ALU*) along with its associated registers (or storage areas) and control unit (Fig. 2.2). The ALU performs all the arithmetic and logical operations of the computer with special registers, called accumulators, providing the input data for the ALU and receiving the results of its operations. In addition to the arithmetic and logic functions, the ALU provides the basic *shift* and *rotate* functions. These functions, illustrated in Fig. 2.3, move the results of the ALU left or right by one or more positions.

The *status* or *condition code* register is a special register that stores the

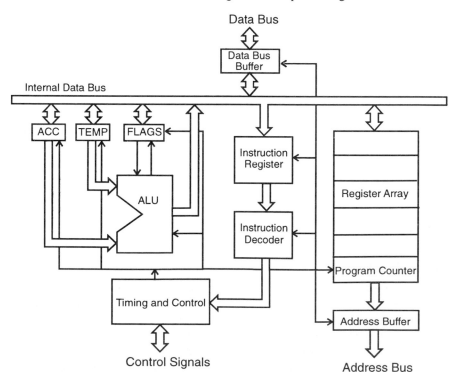

Fig. 2.2 Block diagram of a typical CPU

internal status of the ALU. This register consists of a number of single-bit stores called *flags*. The flags reflect conditions within the ALU such as when the result of an operation is zero, or when an arithmetic operation gives a value too big to hold in a register (*overflow*). The flags are important as it is often the status of these flags that affect the program flow and allows the CPU to make 'decisions' on which instruction to execute next.

Another very important register is the *program counter* (PC). This register is present in all CPUs and is fundamental to program execution as it contains the address in memory of the next instruction to be executed. Whenever the CPU needs another instruction, it places the contents of the PC onto the address bus, then reads the contents of the data bus into the *instruction decoder* to be processed.

The *stack pointer* (SP) contains the address of the top of the *stack*. The stack is crucial in the operation of the computer since it is the mechanism by which the processor 'remembers' its state when it needs to break off from one operation and start another, then return to the original process. Formally the stack is a LIFO (last-in, first-out) structure. It can be visualised as being like a pile of papers: the first item to be put on the pile will always be at the bottom and will be the last to be retrieved, conversely the most recent item put on the top of the pile will be the first to be retrieved.

The final element of Fig. 2.2 to be discussed is the *index register*. This register is used in calculating addresses of data that are needed by the CPU. Typically, this register contains a displacement, which is automatically added to a base value when forming an address. In this way, data within a block can easily be accessed; it is the basic operation involved in creating arrays. Not all CPUs have dedicated index registers and it is often the case that the general-purpose registers can be used instead.

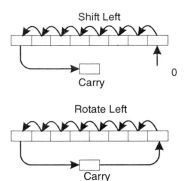

Fig. 2.3 The shift and rotate functions

Input/Output

Input and output, although not fundamental to the operation of the computer, are really its *raison d'être*! There are many types of I/O device: as I sit here typing I am using a keyboard, many of the formatting operations require the use of a mouse, I am looking at the results on a VDU, and the text I am working on is stored on a hard disk. But by and large, we do not have to concern ourselves with how these complex operations are achieved, just some of the fundamentals. I/O devices can be thought of as special memory locations: you can store information at I/O locations and read data from them like any other part of memory. The only difference is that the bits and bytes stored in these locations have some discernible affect on the outside world, and the outside world can affect what information is read from those locations. We will investigate this much further in Chapter 3.

2.2 CPU architectures

The architecture of a CPU is the layout of the registers, ALUs, and control circuitry within the chip. Each type of CPU has its own architecture, and there are generally as many architectures as there are types of CPU. However, the basic way of describing CPU architecture is the complexity of its instruction set. The instruction set of a CPU are the basic operations that it understands: these operations are things such as 'store register A in the

Clock cycles: we saw in Chapter 1 how a *clock* is used to control how some logic elements, such as flip-flops, work. The same principle is used in larger, more complex, circuits. The clock in a computer is the basic signal that initiates transfers between registers, memory and so on and is the basic synchronising source.

memory location pointed to by register B' or 'add register A to the contents of register C'. Some instructions are very basic, such as the store or add instructions; others can be much more complex, such as multiply or moving a block of memory. The complexity of an instruction has a large impact on the amount of time it takes the CPU to execute it; for example, the add instruction takes only 3 or 4 clock cycles to execute, whereas a multiply instruction may take over 150 clock cycles. Things become even worse when complex addressing modes are used. However, analysis of programs shows that these complex instructions are actually used very infrequently, but they contribute a very large part to the electronic complexity of the chip. Consequently, a series of CPUs were developed in which the instruction set was restricted to only the very basic operations, the so-called *Reduced Instruction Set Computer* or *RISC* processor. One of the main benefits of a reduction in the complexity of the instruction set is the ability to *pipeline* instructions. This ability arises from being able to ensure that each instruction in a RISC processor takes the same amount of time to execute, and that the execution of every instruction proceeds in set of well-defined steps. Hence it is possible to start processing the next instruction in sequence while the current instruction is part way through being executed. By arranging for there to be a number of these *pipelines*, it can be possible that the processor will finish an instruction on almost every clock cycle, so vastly speeding up the throughput of the system. Obviously, the presence of branch instructions in the code will mean that the pipelines will break occasionally, but much effort has been put into developing *branch-prediction* and *look-ahead* units in CPUs to reduce their significance.

Another consequence of using a RISC architecture is that caching becomes very efficient; the combination of branch prediction and look ahead vastly increases the efficacy of cache. Indeed all RISC CPUs have on-chip caches of varying size.

This is not to say, however, that *complex instruction set computer*, or *CISC*, processors no longer exist. There are many fields where CISC architectures are important, especially where a computer primarily does one task. Examples of such are database servers: take for example a large company with a stock database, there may be many millions of database queries and modifications (or 'transactions') every second, so to attain the best possible throughput of transactions, highly specialised machine instructions are designed into the processor which aid in the processing of transactions.

Not all manufacturers have polarised into RISC or CISC. Some, such as Intel with their Pentium series, have developed hybrid processors. These are processors that are by and large RISC, but have a number of CISC type instructions. In the Pentium, especially the MMX series, these CISC instructions are present to help the processor, and programmer, to perform multi-media operations (such as displaying and moving images). The problems of introducing these complex instructions into a RISC architecture are seen to be outweighed by the advantages that the presence of these instructions bring.

From a computational point of view, the most important CISC instructions are those that involve floating-point (FP) operations. Many of the time-consuming parts of a scientific program are those that perform calculations on real numbers (as opposed to integers). These FP operations can be performed using integer style instructions with some pre- and post-processing to make sure the calculations are accurate. However, the operations are much faster if there is a dedicated set of instructions to perform them. Indeed there are often FP *co-processors* available to provide this functionality for CPUs which do not have it built-in. Nevertheless, even dedicated FP processors can take some time to calculate complex functions such as logarithms or square roots and consequently the main CPU may spend long periods waiting for the results of the calculations. Modern systems get around this idle time by arranging for the FP unit to be able to work independently and for the main CPU to start a calculation in the FP unit and then resume other operations until the FP result is ready. In this way the overall performance of the CPU is maintained.

2.3 Classes of computers

Historically computers have been divided into different classes according to their power, capabilities and designed use. These classifications are, going from smallest to largest: micro-controllers, microprocessors, microcomputers (PCs), minicomputers, mainframes and supercomputers. Over recent years, a further class, workstations, has been added between microcomputers and minicomputers. However these historical boundaries have been getting very fuzzy recently. It is now possible to have what are essentially minicomputers with the power and architecture of supercomputers; many PCs are now called 'workstations' and have considerably more power than mainframes of only a decade ago. Nevertheless, it is still instructive to examine what are the similarities and differences between these various classes of computers.

Micro-controllers

These days micro-controllers are everywhere: in your washing machine, car, domestic heating system. They can be found in almost any piece of electrical equipment that has more than a simple on-off function! In general micro-controllers are very simple computers that perform one dedicated task, this task being built into the CPU at manufacture and usually not alterable. They are computers though and possess a CPU, memory and I/O functions, albeit at a much lower level than a 'normal' computer – the CPU may only be 4-bit processor and there may only be a few bytes of RAM – but it does follow a predetermined sequence of instructions, and the flow of the program is controlled by external input.

Microprocessors

This is one of the terms in computing which has acquired a multiplicity of meanings. The original concept of a microprocessor is the CPU in a computer, although now it has also come to mean a computer with limited capabilities or one that is used to implement complex control functions. Examples of microprocessor controlled equipment are printers, modems and calculators – the CPUs may be quite powerful and fast, and often have substantial amounts of RAM, but they are still dedicated to doing one task, but that task can be changed, or reprogrammed, after manufacture.

The term 'PC' will be used for all IBM PC compatible computers. Similarly, the term 'Mac' will be used for the Apple Macintosh and its clones.

Microcomputers

Microprocessor based computers that are not dedicated to specific tasks are often called microcomputers. This class of computers is by far the biggest in terms of numbers. It includes all the 'home' computers as well as the vast majority of desktop and 'business' computers. The largest group of microcomputers are the Personal Computers (PCs) manufactured originally by IBM and Apple Corp., although now imitated (or cloned) by many other manufacturers.

Microcomputers are characterised by their ability to perform different tasks according to the program (or software) they are currently executing, and the ability to rapidly change between different programs without the need to change any embedded software. A further feature of microcomputers, which is not usually found in microprocessors, is some form of *mass storage* device. These devices enable the long-term storage of large quantities of information; this information may be either data generated by the machine, or programs for it to execute.

This ability to switch between different tasks has to be controlled in some way, and most microcomputers have some form of supervisory program performing this rôle called the *operating system* (OS). There are many different operating systems available for microcomputers; you will most likely come across MS-DOS or WinNT for PCs or MacOS for Macs, but there are also OS's such as Linux for both PCs and Macs or FreeBSD or O/S2 for PCs that you may encounter. There will no doubt be many more OS's written in the future, but it is likely that they will, at least in the near future, be based in some way on one of the ones mentioned above!

As indicated before, microcomputers are the most common class of computers, indeed most of the software and hardware described in this book will be based on microcomputers, and more specifically PCs. This does not mean that it is not applicable to other machines, just that it is the most likely scenario you will come across.

Workstations

Some consider workstations to be 'grown-up' versions of microcomputers; others consider them to be 'stripped-down' minicomputers! Whichever is the case, workstations can generally be said to be powerful personal computers. Often they have very good graphics capabilities, a large amount of memory and are connected to a network. The software run on workstations tends to be some form of *multitasking* OS, enabling the machine to run more than one process at a time; the individual processes are often graphically based and are controlled through a *graphical user interface* (GUI). However, the term workstation is more of a description of the intended use of the machine than an absolute name for a type of machine: one person may use a type and model of machine as a personal computer/workstation, whilst another may use the same machine as a server/minicomputer.

At the low-end of the workstation range are the powerful PCs and Macs, whilst at the high-end are powerful machines from manufacturers such as Silicon Graphics, Sun, Hewlett Packard, IBM and so on. In-between is a large range of machines from a number of manufacturers.

Minicomputers

Most of the 'large' computers chemists encounter will come in this class. Minicomputers are medium sized multi-user machines that are almost invariably housed in a single box or cabinet. They will have one or more CPUs working together and will have moderate amounts of both RAM and mass storage. The OS will probably be some form of *Unix*, but might also be VMS or some other proprietary OS. Often these computers are *servers*, that is, machines that provide information for other computers to use.

Minicomputers will usually have compilers or interpreters for a number of different high-level languages to enable the user to write their own programs, as well as many other utilities for performing common tasks. In general the user will never need to know, or care, what sort of CPU the machine uses, as they do not need to interact directly with the hardware. Typically the manufacturers of these machines are Sun, Silicon Graphics or Hewlett Packard amongst others.

Mainframes

The term 'mainframe' comes from the days when computers occupied whole rooms and there were many different cabinets, or frames, containing electronics that performed discrete functions; the 'mainframe' was the set of cabinets containing the central processor along with the main memory of these physically large machines. Now the term 'mainframe' refers to any large computer, although more recently it has come to be used mainly for very large servers and for machines that the user doesn't interact directly with. A mainframe will probably have multiple processors or consist of a *cluster* of tightly coupled machines; the CPUs will be proprietary and have a CISC architecture (although RISC based mainframes are starting to appear). The whole machine, especially the CPU architecture, will often be tuned for a specific application, such as a database server.

Supercomputers

Supercomputers are the glamour end of the market! These machines are very large often with hundreds of linked high-speed processors and vast quantities of memory. They are typically used for very large calculations in meteorology, materials science or astronomy and are tuned for doing large floating point calculations. The calculations are often performed across many of the processors at once: such *parallel processing* is very efficient when large arrays are being manipulated. By far the most famous supercomputers are those of Cray Research, although other manufacturers such as Silicon Graphics, CDC, Fujitsu, NEC, and Hitachi also produce supercomputer class machines.

Surprisingly, supercomputers are often interactive machines and appear to the user to be like a powerful minicomputer. However interactive use is not efficient on these machines, so they are often linked with a minicomputer *front-end* which provides the basic user interface, and only the large computing jobs are actually run on the supercomputer.

2.4 Examples of hardware

As examples of the internal architecture of different computers we will take a detailed look at two very different machines. The first is, inevitably, the IBM PC compatible, the second is the Silicon Graphics Origin2000 series. The IBM PC is a personal microcomputer and can be found in almost any laboratory: it is used to control experiments, as a database, a word processor, a terminal and much, much more; the SG Origin is a minicomputer with many supercomputer features and in general it is a department or campus based machine. These particular machines have been chosen for no other reason than that they are the ones that the author is most familiar with.

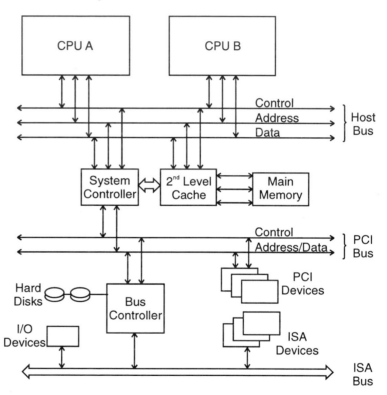

Fig. 2.4 Simplified block diagram of a PC

The IBM PC

A simplified block diagram of the internal structure of a PC is shown in Fig. 2.4. As can be seen it is considerably more complicated than that shown in Fig. 2.1. The system shown here happens to be a dual CPU one; however that in itself is not important, as the presence of a second CPU has very little bearing on the hardware present, it is mainly a matter for the software.

The presence of three separate bus systems is of note and is largely due to the problems of interfacing with external hardware and maintaining compatibility with legacy devices. When the first IBM PCs were designed, the CPU operated at a clock rate of 4.75 MHz, and naturally all the rest of the circuitry followed this speed, including the external bus (called the ISA bus). Shortly thereafter the speed of the CPU increased to 8 MHz, and the ISA bus followed without too many problems (although some devices designed for the original machine did stop working). When the second generation of IBM

PCs (the PC AT) were designed, they used an upgraded CPU (the 80286) which was capable of working at much higher speeds (up to 20 MHz); however it was not possible to reflect this clock speed in the ISA bus: the specifications of the bus stipulate a maximum clock speed of 8 MHz. Consequently, it was necessary to slow the whole machine down, whenever a device on the ISA bus was accessed. This slowing down was achieved by inserting *wait states* into the CPU clock – effectively making the CPU miss one or two clock cycles while the ISA bus caught up. Unfortunately, as more and more programs relied on I/O for their operation (such as GUIs relying on access to the video systems), it was found that the CPU was spending much of its time in wait states. So, to alleviate this problem, it was necessary to have another bus system, in addition to the ISA bus, which allowed access to peripherals at the full speed of the processor. A number of proprietary systems were developed, the *local bus system* being the most widely used of the early versions. However there were a number of 'flavours' of local bus system, and none became an industry standard. Eventually though, an existing industry standard, the 33 MHz PCI bus, was introduced into PCs and became accepted.

This is not the end of the story for bus speeds. Many modern CPUs have stated speeds of 400 MHz or more. However, this does not mean that all the buses operate at these speeds. The speed stated is the internal bus speed of the chip, *i.e.* the speed at which the chip communicates between the ALU, registers, and, very importantly, the internal primary cache; when information is needed from an external source to the CPU, the bus speed is dropped to between 33 and 100 MHz. This means that there is still a need to provide interface circuitry between the host bus (running at, say 66 MHz) and the PCI bus (running at 33 MHz). Most PCs still have a number of ISA bus slots for slower peripherals: the ISA bus is now usually a 'device' on the PCI bus. In the future there will undoubtedly be enhancements and modifications to this picture: already the PCI2 bus running at 66 MHz is being introduced into workstation class machines, and CPUs operating at 1 GHz are being developed.

The bus each peripheral is attached to is largely dependent on the speed at which it operates: high bandwidth devices, such as video cards, disk controllers and high speed network cards, will be attached to the PCI bus, whereas slower devices, such as the serial and parallel ports, keyboard, mouse, audio and network cards will use the ISA bus. Most devices used for interfacing the PC to experiments will also be attached to the ISA bus as usually the rate of exchange of information is quite low.

The Silicon Graphics Origin2000

The Origin2000 is a multiple processor computer consisting of from 2 to 256 MIPS processors and up to 512 GB of memory. As would be expected, it is a very complex machine, and it will only be possible to scratch at its surface here. Nevertheless, it is instructive to examine some of the features of its design so that the basis of the advice on efficient programming given in Chapter 5 will be more easily understood.

The main feature of the system is the method by which groups of these processors intercommunicate. The basic system building block is a *node board*, each node board contains two processors and their associated cache (usually 4 MB), along with some main memory. The node board also contains a *hub* that controls the buses connecting the processors and memory. Each pair of node boards is then connected to a *router*; it is the router which controls how each of the node boards (and thus ultimately each processor) communicates in the system. All the routers in a system are interconnected (*via* the *CrayLink Interconnect*) in a manner that allows the most efficient interchange of data between the processors. A total system then consists of up to 16 modules, with each module containing 2 routers, 4 node boards and 8 processors.

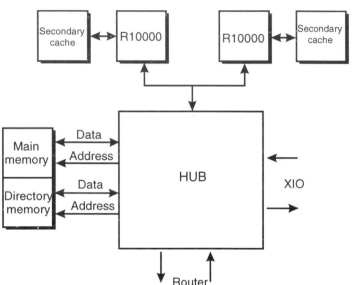

Fig. 2.5. Block diagram of the Origin2000 node board (© see footnote on page 24)

Figure 2.5 shows a simplified block diagram of a node board. The XIO port is the bus used to connect with peripherals such as disks or networks; it can be thought of as being equivalent to the I/O bus in PCs, although it is considerably faster. Each XIO system consists of an 8-port crossbow interconnect system, with 2 of the ports on the crossbow connected to two of the node boards in a module (either boards 1 & 3 or 2 & 4), with the other 6 ports being connected to individual XIO devices. The crossbow is simply a way of being able to connect any two ports on the device together: hence a node board can talk to any of the 6 XIO devices, and each of the XIO devices can talk to each other. Fig. 2.6 shows the interconnections between the various components in a module.

Each router board can connect up to six devices. Two of those devices are the node boards, and a further one is the other router board in the same

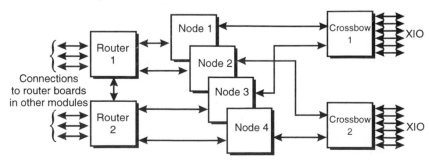

Fig. 2.6. Block diagram of an Origin2000 module (© see footnote on page 24)

module. This leaves three ports available for connection to other routers. In general, there are two modules in each rack, and so one of the external connections of each router is connected to the corresponding one in the other module in the rack. In this way, the individual routers in a rack are connected in a square and there arc at least two routes between any two routers (Fig. 2.7a). When there is more than one rack, each of the routers within each rack is connected in this square topology, and then each router is connected to the corresponding one in the rack next to it, forming a cube arrangement (Fig. 2.7b). If there are 3 or 4 racks, then a similar style of expansion is made, *i.e.* each pair of racks is connected in the cube arrangement, then each vertex on the cube is connected to the corresponding one on the other cube, producing a 'hypercube' topology (Fig. 2.7c). Finally, if there are more than 4 racks (or 64 CPUs), then each vertex of the cubes formed from each pair of racks (*i.e.* every router) is connected to a special device called a 'SGI Meta Router' (Fig. 2.7d). This is a separate cabinet containing a network of routers and that allows all the node boards to talk to each other. Currently up to 32 modules (256 CPUs) can be interconnected using the Meta Router.

In designing a large multiple processor system such as the Origin2000, many factors need to be taken into account. For instance, the fact that the main memory is distributed amongst all the node boards means that, unless precautions have been taken, the memory that a processor is accessing is likely to be on a different board to that processor, and no matter how fast the interconnections are, it will take longer to access that memory than if it were in the same location. A large processor cache tends to alleviate this problem (since with a large cache, the original memory locations need not be accessed as often), but it is still a factor that needs to be taken into account when the system software is being designed. A further, potentially more intractable problem is that of *cache coherency*. Cache coherency is making sure that the

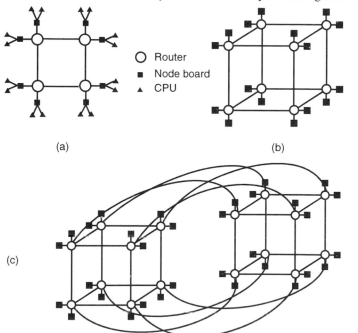

○ Router
■ Node board
▲ CPU

(a)

(b)

(c)

Fig. 2.7. Router connection topologies for the Silicon Graphics Origin2000.
(a)   8 node boards in a single rack (*i.e.* up to 16 CPUs)
(b)   16 node boards (up to 32 CPUs)
(c)   32 node boards.
For clarity the individual CPUs are only shown in Fig 2.7a, each node board implicitly has two CPUs connected to it.
(© see footnote on page 24)

Fig 2.7d. The router interconnections in a 128 CPU Silicon Graphics Origin2000 system. Each cube of routers is connected to the others *via* the SGI Meta Router.
(© see footnote)

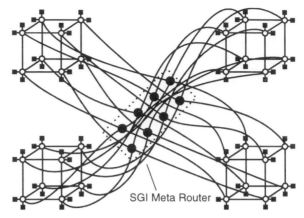

SGI Meta Router

value that is currently held in a cache correctly represents the value that is in main memory. In a single processor system, this is no problem: in general, only the processor can change memory and so the value in the cache must be the same as in main memory. However, in a multiple processor system, any of the processors may alter a memory location, and so steps must be taken to ensure that when one processor does change a value, then all the other processors are told that any value they have in their cache corresponding to that location is now wrong. There are various schemes to ensure cache coherency; the brute force method is to clear the memory location from cache, whenever another processor writes to that location, thus forcing a re-read the next time the memory is needed. The more subtle methods include broadcasting the new value to all the processors so that the correct value can be put in the cache. The Origin2000 uses a hybrid technique: the memory subsystem keeps a record of each processor that is using a particular memory location (this is the purpose of the *directory memory* in Fig. 2.5), when a processor needs to write to a location, it informs the memory subsystem which in turn tells all the processors that are using that location to invalidate that particular cache entry. When a value from that location is needed, the processor sends a request to the memory, which passes it on to the processor that invalidated the cache, this processor then sends back the correct value to both the processor that originated the request and the memory system. In this way the integrity of the cache is maintained without the expense of large numbers of messages passing between CPUs in order to maintain the coherency.

Figures 2.5, 2.6, 2.7a-d: Illustration concept courtesy of Silicon Graphics Inc. © 1999 Silicon Graphics Inc. Used by permission. All Rights reserved.

3 Interfacing computers to experiments

In order for computers to be able to process data, the data must be first entered into the machine. Where the primary purpose of the computer is analysis of end results, then the data is often entered by hand – the experimenter *is* the interface. But when the purpose of the computer is analysis of primary data, such as counting events, measuring voltages, and so on, the human interface looses its attraction – it is slow, error prone, and lacks concentration. This chapter provides an introduction to the techniques of both getting experimental information into a computer and causing a computer to control the experiment.

Because of limitations of space (and knowledge), it will be necessary to restrict the examples to those involving the PC. The techniques described will be applicable to other computers; it will only be the interface between the computer hardware and the circuits described here which would change.

3.1 Introduction to interfacing techniques

In reality there are only a limited number of things that need to be measured or controlled in an experimental situation: voltages need to be measured or generated, events need to be counted or generated, and times need measuring. Any other experimental parameter can be converted into one of these basic functions. For instance, temperature or pressure is very easily converted into a voltage, indeed most temperature or pressure sensors work by generating voltages, or their control unit has an output which can be used. Any output that is displayed on, say, a chart recorder is already a voltage. Devices such as Geiger counters, whose output is a series of pulses, can be interfaced either by taking the output of their rate meter as a voltage or by counting the pulses directly.

Perhaps the most difficult quantity to measure electronically is distance. Small distances (< ~2 cm) can be measured accurately using differential transformers (see Wayne), moderate distances (up to ~1 m) are measured using optical linear position sensors, longer distances (up to 10 m) can be achieved with ultra-sound ranging, and for the longest distances laser ranging is used. But they all have one thing in common, the output is finally delivered as a voltage that is proportional to the distance being measured.

The device used to 'measure' a voltage is called an *analogue to digital converter* or ADC; similarly, the complimentary device used to generate a specific voltage is called a *digital to analogue converter* or DAC. The detailed operation of both ADCs and DACs is described below; here it is sufficient to say that an ADC takes a voltage and converts it into binary data proportional to that voltage (and *vice versa* for DACs). In a computer system, the output of the ADC (or input of the DAC) is designed so that it looks like one or more memory locations on the microprocessor bus. Thus, in order to read the voltage present on the input of an ADC, all that is needed is to perform a read on the corresponding memory location.

Counters are also arranged so that they appear as one or more memory locations on the bus. Generally though, counters are both readable and writeable – they are readable, obviously, so that the current count can be read, but they are also writeable so that a preset value (usually zero) can be loaded into the counter before it starts counting. Often the ability to start and stop the counter is provided, either through programming or by an external signal. This start/stop function is important when the counter is used as a rate meter, since it is necessary to count the number of events over a fixed period of time. As an extension of this, using an external signal to start and stop the counter, and providing the counter with an accurately known frequency source as its input, allows the time between the start and stop events to be measured. For instance, if a counter is provided with a 1 MHz input frequency, every increment to the count corresponds to a time increment of 1 µs; hence if there is an interval of 34 µs between the two events, the counter will count to 34 or 00010010 in binary.

The final class of device that will be looked at here is the *parallel I/O* or PIO device. The PIO allows a single bit (or groups of bits) to be output or input. The importance of such functionality is in being able to sense whether a signal is on or off, and in being able to control on/off devices such as relays, valves, motors, lamps, *etc.* A Schmitt trigger device usually buffers the input signals before connecting to the PIO; similarly the output signals are buffered before being used to drive the external device.

Digital to analogue converters

The principle of a digital to analogue converter (DAC) is simple: the idea is that the device produces at its output a voltage proportional to the binary data at its input. One way of achieving this is to arrange for a specific point on a voltage divider chain to be selected for each combination of inputs. This is feasible for small word lengths – a 3-bit word would only require 8 tap points on the divider chain (see Fig. 3.1) – but with the more common word lengths of 10 or 12 bits, the number of tap points becomes unreasonably large. A further problem with such voltage divider chains is that as soon as any current is drawn from the output, the progression of voltages is unbalanced and the output becomes non-linear.

Measuring time using an external counter and frequency source is much more accurate than using the CPU clock – the CPU clock is often slowed down or stretched when peripherals are accessed. In addition, if an interrupt occurs, the timing will be inaccurate.

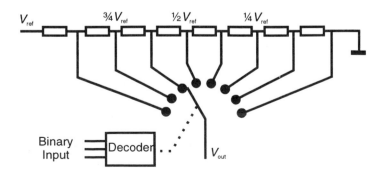

Fig. 3.1 A voltage divider chain as a 3-bit digital to analogue converter. The output voltage, V_{out}, is $V_{ref}.N/8$, where N is the decimal equivalent of the binary input.

The solution is to perform some sort of summation. Voltage addition is not straightforward, so usually current summation is used and is illustrated in Fig. 3.2 for a 4-bit word. Here, each constant current source is arranged to produce twice the current of its neighbour in a binary progression; the total output current is thus proportional to the input binary data. The output voltage is then developed by passing this current through a resistor.

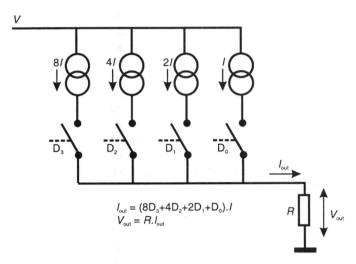

$$I_{out} = (8D_3+4D_2+2D_1+D_0).I$$
$$V_{out} = R.I_{out}$$

Fig. 3.2 A 4-bit current summation DAC. The voltage developed across the resistor is dependant on the state of the four switches.

The use of constant current sources is useful in illustrating the principle of how a current summation DAC works. However, in practice, accurate, stable constant current sources are difficult to fabricate. The most common form of current summation DAC are those which use a current switching R-$2R$ resistance ladder. The basic idea of these devices is that when two resistors of equal value are placed in parallel, the same current flows through each (Fig. 3.3); further, if each resistor has a value of $2R$, the total resistance is R. If then this pair of resistors is used, along with another resistor of value R, in one arm of another parallel pair, then the same equal current division will be achieved, but the current in each of the original pair of resistors will be ¼ the total current in the circuit (Fig. 3.4). It is now obvious that cascading the parallel resistance circuits in this way causes the current in each arm to be halved each time. Thus we have our binary progression of currents.

A big advantage of the R-$2R$ network is that only resistive components are present. Consequently it easy to adjust, or *trim*, the network, if necessary, at the fabrication stage in order to ensure a linear response.

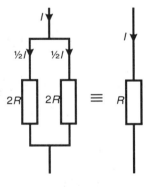

Fig. 3.3 Two parallel, equal value, resistors have the same current flowing through them

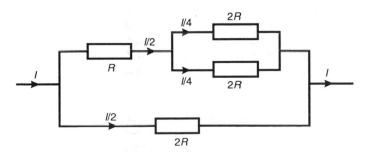

Fig. 3.4 *R*-*2R* resistor network used to create a binary progression of currents

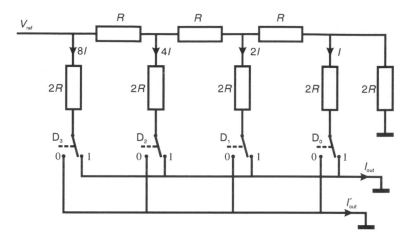

Fig. 3.5 A current switching network. The output current is given by the same formula as in Fig. 3.2. I'_{out} is the inverse of I_{out}, such that $I_{out}+I'_{out} = 15I$. The current, I, is given by $V_{ref}/16R$.

The details of operational amplifiers can be found in either Wayne or Horowitz and Hill.

In a real circuit, each current flow is switched to one of two places depending on the binary input data (Fig. 3.5); the current flows in each output are thus either proportional to, or inversely proportional to, the input binary data. However, it is important that each of the outputs is maintained as close to zero potential as possible, otherwise the current flow in the circuit will become unbalanced and will no longer be linear, consequently it is not possible to just pass the current through a resistor in order to develop the required output voltage. In most applications, the output current is converted to a voltage using an operational amplifier (Fig. 3.6).

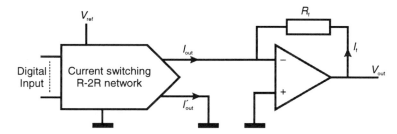

Fig. 3.6 An operational amplifier used to convert the output current of the R-$2R$ network into a voltage

The inverting input of the operational amplifier is a "virtual ground" created by the feedback current from the output through R_f. This feedback current is equal in magnitude, but opposite in sign, to I_{out}. Thus the voltage at the output of the operational amplifier (V_{out}) must be $-I_{out}/R_f$. If R_f is chosen to be the same as the value of R in the R-$2R$ ladder, then it can be shown that V_{out} is $NV_{ref}/2^n$, where N is the decimal equivalent of the binary input data and n is the number of bits in the DAC.

In all these circuits the stability of the output voltage depends on the stability of the current, I. This current, in turn, depends solely on the voltage across the resistor network. Consequently, the excitation or *reference* voltage is usually provided by an extremely stable source. The reference voltage is also often chosen to be a convenient value; for instance, a 10-bit converter might have a reference voltage of 1.024V, resulting in a resolution of 0.001V and an output voltage which is one thousandth of the decimal value of the input.

Analogue to digital converters

There are two main types of ADCs: flash converters and successive approximation converters. The simplest in concept is the flash converter. In these devices the input signal (V_{in}) is fed to a series of comparators, each comparator's reference voltage (V_r) is derived from a resistance divider chain, the top of the chain being provided with a stable reference voltage (V_{ref}). If all the resistors in the divider chain are of equal value, then the first comparator from the top of the chain whose output is high (i.e. $V_r > V_{in}$) is proportional to the input voltage. In the ADC, the output of all the comparators are passed to an encoder, the binary output of which depends on the highest numbered input that is high. Thus, the output of the ADC is a binary number representing the input voltage. Fig. 3.7 shows a simplified flash ADC. The advantage of this type of ADC is that its operation is very simple and rapid. However, if high accuracy is required, then the number of comparators increases exponentially with resolution: for 4-bit accuracy 16 comparators are required, 8-bit needs 256 comparators, 10-bit needs 1024 and so on. As the number of comparators increases it become more difficult to fabricate all parts of the circuit with identical characteristics and so it becomes more difficult to attain the required accuracy and linearity.

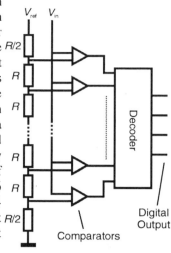

Fig. 3.7 A flash analogue to digital converter

The second type of ADC is the successive approximation converter. In this device, the input voltage is first compared to half the reference voltage, if V_{in} is less than $\frac{1}{2}V_{ref}$, it is then compared to $\frac{1}{4}V_{ref}$; if it is higher, it is next compared to $\frac{3}{4}V_{ref}$, and so on. The true value of V_{in} is thus 'zeroed-in' on. A simplified diagram of such an ADC is shown in Fig. 3.8. The circuit is complex, but can easily be divided into three sections: the successive approximation register (SAR), the DAC and the comparator. The detailed circuitry of the SAR is beyond the scope of this book, but essentially it is a combination of shift register and latches. The operation is as follows: on the rising edge of each clock pulse, after the application of a start signal, an output, starting at the MSB, goes high, on the falling edge of the clock, that output takes on the same state as that present at the D input, on the next clock rising edge, the next output goes high and so on. The sequence repeats until

MSB = Most Significant Bit
LSB = Least Significant Bit

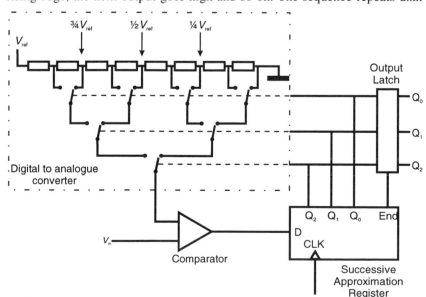

Fig. 3.8 A simplified diagram of the successive approximation analogue to digital converter

the last bit has been determined and on the next clock pulse the *end of conversion* (EOC) signal goes high. The sequence of events in the ADC as a whole is as follows:

1. When the start signal goes high, the SAR is cleared and Q2 is taken high. The 3-bit DAC thus gives at its output a voltage equivalent to half the reference voltage.

2. The output of the comparator gives a logic zero if V_{in} is lower than V_r ($= \frac{1}{2}V_{ref}$), and a logic one if it is higher, and, since the output of the comparator is connected to the D input of the SAR, on the next clock edge, Q2 will take on the same state as the comparator output. Thus the first *approximation* is complete.

3. On the next clock edge, Q1 is taken high and the cycle is repeated: each output of the SAR takes on the value of the comparator at the time.

A typical sequence of events for a 4-bit ADC is shown in Fig. 3.9. In this diagram the state of each output on each clock edge is shown, along with the output voltage of the DAC. The shaded areas represent the range of voltages eliminated by the ADC on each approximation, showing how the true value of the voltage is zeroed-in on. Note also that the final converted voltage is not exactly the same as the input voltage. This inaccuracy is because the resolution of the ADC is not high enough to accurately represent the true voltage.

The resolution of an ADC is determined by the number of output bits. A 4-bit converter will only have 16 discrete output states, so if the reference voltage is 1.6V, the resolution will be 0.1V.

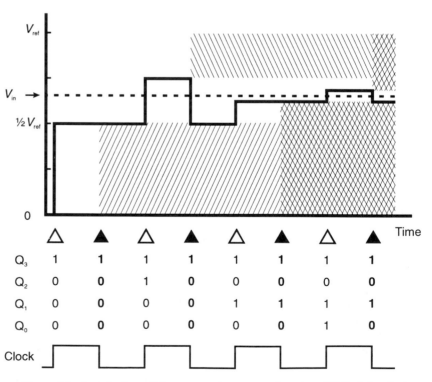

Fig. 3.9 The operation of a successive approximation ADC. The open triangles show the point at which an approximation is made, the filled triangles show when a 'decision' is taken. The final output of the ADC is 1010, indicating that the input voltage is approximately $10/16 \; V_{ref}$.

From this description of the successive approximation ADC, it is obvious that the conversion is not instantaneous like the flash ADC described previously. The conversion time for an *n*-bit ADC is usually around *n*+1 clock cycles: typically this means around 100μs for an 8-bit conversion compared to <1μs for a flash ADC. Nevertheless, the advantages of the successive approximation ADC often outweigh their slowness. For instance, they are usually more accurate than flash converters – accuracy in this case usually means linearity, *i.e.* a change in 1 bit represents the same voltage no matter where on the scale that bit change occurs: in flash converters, this linearity is a function of both the resistor divider chain and how identical each comparator is, and so may depend on many components; whereas in the successive approximation converter, the linearity is entirely dependent on the DAC and the resistor divider chain therein.

Sample and hold

One big disadvantage of a successive approximation ADC is that in order to obtain an accurate conversion, the input voltage must be held constant throughout the whole of the conversion period. In an experimental situation, this is not often, if ever, possible. The solution is to use a *sample and hold* (S/H) device. These devices use the ability of a capacitor to store charge in order to keep the voltage at the input to the ADC constant. A simplified circuit of a S/H device is shown in Fig. 3.10 and consists of an input buffer, an electronic switch, a storage capacitor and an output buffer. The storage capacitor is charged to the same voltage as V_{in} when the switch is closed; when the switch is opened, since the input impedance of the output buffer is high, the voltage on the capacitor remains constant, and hence the output voltage, V_{out}, remains constant. Thus, V_{out} takes on the same voltage as V_{in} every time the switch is closed, and holds that voltage until the next time the switch is operated. The length of time that the S/H will hold a voltage for is dependent on the value of the storage capacitor – the higher the value, the longer it will hold the charge for – but also, the higher the value, the longer it will take to attain a stable charge, and so fast changes will not be able to be followed accurately. Typically, storage capacitors of about 0.01 μF are used, enabling acquisition times of about 50 μs to be achieved with drift rates of about 0.5 mV s^{-1}.

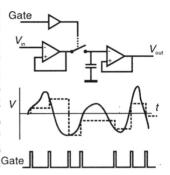

Fig. 3.10 Simplified sample and hold. The output (dashed line) takes on the same voltage as the input (solid line) on each gate pulse.

PIO devices

The ability to detect if an external device is on or off, or indeed to switch an external device on or off, is of paramount importance in an experimental situation. Although it is possible to construct individual on/off circuits, it is usually the case that a number of such elements are grouped together in one device; these devices are called parallel input/output (PIO) devices. These devices are found in just about every computer: it is the basic element that forms the parallel printer port. Indeed the parallel printer port can often be subverted into acting as an external experimental controller in some circumstances, but for simplicity, we will not be taking that approach here.

The individual I/O lines of a PIO device are often grouped into convenient sets. Most often these sets, or ports, are 8 bits wide. There are numerous PIO devices available, and each has their own individual features. For instance, with some, it may be possible to program each individual bit to be either an input or an output, some it may only be possible to program *ports* to be input or output, some may be able to act as both input and output simultaneously; some may have Schmitt trigger inputs, whereas others may have high current outputs. The device we will concentrate on here is the Intel 8255; this may not be the most powerful PIO available, but it is very versatile, and is easy to understand and program.

The 8255 is a general purpose I/O device designed for use with microprocessors. It has 24 I/O lines that may be used in a number of different ways. These 24 lines are divided into three basic 8-bit ports: A, B, and C. Ports A and B can be programmed as either input or output whereas port C can be programmed in groups of 4 bits to be either input or output, or act as control and status signal lines for ports A and B. A block diagram of the device is shown in Fig. 3.11. The two address lines A_0 and A_1 are used in conjunction with the chip select ($\overline{CS}$) and the read and write ($\overline{RD}$ and $\overline{WR}$) lines to access a number of registers in the chip. These registers and their operation are shown in Table 3.1.

Fig. 3.11 Block diagram of the 8255

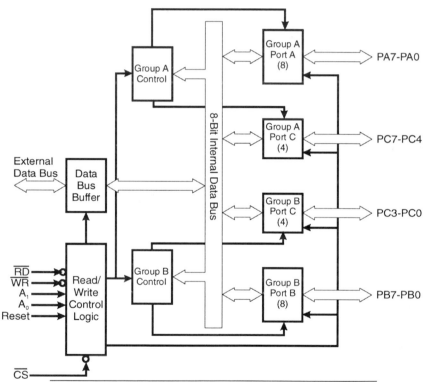

Table 3.1 The basic operation of the 8255. × = "don't care"

A_1	A_0	$\overline{RD}$	$\overline{WR}$	$\overline{CS}$	Operation
0	0	0	1	0	Port A $\Rightarrow$ Data Bus
0	1	0	1	0	Port B $\Rightarrow$ Data Bus
1	0	0	1	0	Port C $\Rightarrow$ Data Bus
1	1	0	1	0	Illegal
0	0	1	0	0	Data Bus $\Rightarrow$ Port A
0	1	1	0	0	Data Bus $\Rightarrow$ Port B
1	0	1	0	0	Data Bus $\Rightarrow$ Port C
1	1	1	0	0	Data Bus $\Rightarrow$ Control
×	×	×	×	1	Data Bus disabled
×	×	1	1	0	Data Bus disabled

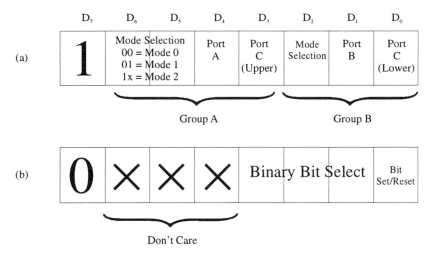

Fig. 3.12. The control register of the 8255

(a) Setting the mode and data flow direction. A '1' in the bit representing each port sets that port to be an input, whilst a '0' sets it to be an output.

(b) Single bit set/reset of Port C. The bit is selected using D_3-D_1, and is set ('1') or reset ('0') according to D_0.

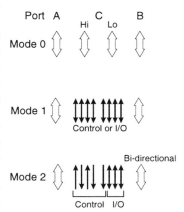

Fig. 3.13 Basic mode definitions in the 8255

The control register is an 8-bit write only register that both determines how each port behaves and allows individual bits of port C to be modified. When bit 7 of this register is 1, the other bits determine the operational mode as shown in Fig. 3.12a. There are three basic modes of operation: Mode 0 is simple I/O giving 24 programmable I/O lines; Mode 1 is used for *strobed* I/O in which port C is used to provide *handshaking* and *status* signals for ports A and B; Mode 2 implements a bi-directional bus I/O system on port A along with five of the lines of port C. These modes are shown in Fig. 3.13.

When bit 7 of the control register is 0, the rest of the bits are used to set or reset individual lines of port C (Fig. 3.12b). This functionality enables individual bits to be set or reset using a single processor function, thus simplifying program design.

The details of all the modes and combinations of modes and their applications are far too complicated to go into in full here. However, if it is ever necessary to program this, or indeed any other, device, then a very careful read of the manufacturers data sheet will provide you with all the necessary programming information.

Pulse generation

It is often necessary to generate pulses, either individually or in sequences, when interfacing to experiments. Obviously, it is simple to generate a single pulse using a PIO: it is only necessary to arrange for an output to be turned on, then a short time later turned off. However, because of the nature of a computer, it is not possible to ensure that the 'short time later' is always the same; in some cases it may not be important, but in many, the length of the pulse, or the time between two pulses, is of paramount importance. The easiest way to overcome this is to use some form of external timing device.

If a pulse needs to be of constant, unvarying, width, then it is only necessary to have the leading edge of an output of the PIO trigger a monostable; the length of the pulse from the PIO is then of no consequence. However, this does mean that the length of the pulse can not be under program control. As alluded to earlier though, it is possible to use counters as timing devices. In order to do this, it is necessary to connect a stable, known frequency source to the clock input of the counter: each count of the counter,

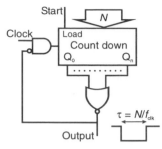

Fig. 3.14 A count-down timer. A *start* pulse loads a value N into the counter. The counter then takes N clock pulses to reach zero again, at which point the clock pulses are inhibited. The *zero* output thus gives a pulse whose output is exactly N clock pulses long.

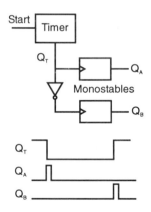

Fig. 3.15 A timer used to create a delay between two pulses.

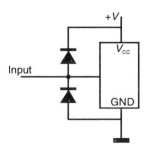

Fig. 3.16 Diodes used to limit, or *clamp*, the input voltage to the supply voltage.

then corresponds to time interval of $1/f_{clk}$ (for instance, a clock frequency of 1 MHz will correspond to a time interval of 1 μs). It is important that this frequency source is independent of the computer clock signals: the computer clock is often stretched or stopped according to which peripherals are being accessed and so it is neither stable nor known!

The counter in this case acts as a 'count-down timer': a number is loaded into the counter, and the counter counts down to zero from this number; the time it takes to reach zero is determined by the number loaded. The pulse output is taken from a 'zero' output of the counter, i.e. the output is high when all the outputs are zero and goes low when a value is loaded into the counter, the output then goes high again when zero is reached. It is also important to ensure that the counter is stopped when the count down has finished, this is usually achieved by using the zero output to inhibit the clock pulses. A typical arrangement is shown in Fig. 3.14.

A similar arrangement can be used to provide a time delay between two pulses: the zero count output is used to negative edge trigger a monostable pulse generator, with the positive edge being used to trigger another monostable (Fig. 3.15).

There are two main factors to consider when using this sort of circuit: the accuracy of the timing, and the period required. In turn, these two factors are dependent on the frequency of clock used and the number of bits in the counter. For instance, an 8-bit counter clocked at 100 kHz will have a resolution of 10 μs but a maximum period of only 2.55 ms. If a longer period is required, either the clock rate can be decreased (a clock of 1 kHz will give a 255 ms maximum period, but with a resolution of only 1 ms), or the size of the counter can be increased (a 16-bit counter clocked at the same rate will have the same resolution, but a maximum period of 655.35 ms). Thus there is a trade off between resolution and maximum period for a given size of counter, the values used depend on the needs of the experiment.

More complex counter/timers are, of course, available, and using these it is possible to build up complicated pulse sequences such as those used in NMR.

3.2 Signal handling techniques

The characteristics of the external signals output from and input to an interface device are often restricted to a small range of values. For instance the outputs of a PIO device are usually TTL level (i.e. low is < 0.8V, high > 3.4V) whereas it may be necessary to provide an output voltage of, say, 12V; similarly, the output current range of a TTL device is very limited, with standard devices only being capable of providing a maximum of 16 mA. Consequently it is often necessary to provide extra circuitry between the interface device and the 'real world'.

Protecting the chips

Although most common interface chips are sensitive to static charge, once they are placed in a circuit, they are very robust. However, one sure way of damaging any sort of chip is to put too high a voltage on one of its inputs. There are various techniques used for protecting a chip from accidental damage in such a way, the simplest of which is *diode clamping* (Fig. 3.16).

Here two diodes are placed such that if the value of the input voltage remains between the supply voltage and ground, both diodes will be reverse biased and will not conduct; however if the voltage goes beyond either of these two limits, one of the diodes will start to conduct, and the input will be clamped to the supply voltage. If the input can not tolerate voltages as high as the supply voltage, such as in some ADCs, then the input can be protected using a zener diode; in this case, the input is, obviously, clamped to the zener voltage (Fig. 3.17).

Sometimes, just clamping the input voltage will not be possible. For instance, voltages much higher than the supply voltage will cause excessive current to flow in the clamping diodes, and, possibly more seriously, the clamping action of the diodes may affect the quantity being measured! Often the addition of a voltage divider on the input, or even just a current limiting resistor, in combination with diode clamping will allow voltages as high a few kV to be measured or sensed. However, there are times when none of these approaches are satisfactory. For instance, if the ground voltages between the measuring and measured circuits are incompatible (*i.e.* it may be necessary to measure a voltage referenced to a voltage other than zero), then some form of isolation between the two circuits is necessary.

For a description of the operation of a zener diode, see Wayne's book.

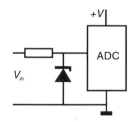

Fig. 3.17 Clamping the input of a sensitive device using a zener diode.

Optical isolators

The most common form of isolation in digital circuits is an optical one. Special devices, called *opto-isolators*, containing a light emitting diode (LED) optically coupled to a photo transistor are the most commonly used. These devices provide electrical insulation between two circuits up to about 3 kV, although the actual degree of isolation is usually determined by the circuit board on which the devices are mounted! A typical arrangement is shown in Fig. 3.18. Here an opto-isolator is used to convert between the −12 V signal level in one part of the circuit and the TTL levels in the next. It should also be remembered that opto-isolators can be used as output devices: Fig. 3.18 could equally well have shown the TTL level device driving the LED and the photo transistor controlling the −12 V signal.

Opto-isolators are not just digital devices: within defined ranges the coupling between the emitter and receiver is linear. Indeed, these devices are often used in high voltage control circuits to isolate the sensing (high voltage) circuits from the controlling (low voltage) ones.

Optical isolation has many other uses. For instance, it may be necessary to control a device in an electrically noisy environment such as near a pulsed laser. Long lengths of cable, even screened cable, inevitably pick up the large current surges in these pulsed devices; this electrical noise often causes many problems in sensitive digital circuits, causing spurious pulses to be detected. The solution is to use optical fibre, instead of ordinary cable, between the laser and the digital circuit. Such an arrangement is shown in Fig. 3.19. In this particular example, two lasers are being triggered from one pulse source with a delay between the pulses – if optical isolation were not used, then there would be a risk that the noise pickup from the first laser firing would cause the second laser to fire prematurely.

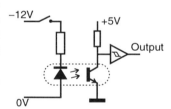

Fig. 3.18 An opto-isolator used as an interface between two incompatible voltage levels.

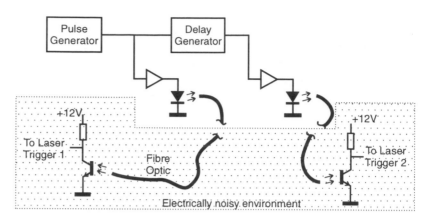

Fig. 3.19 Fibre optics used to trigger lasers. The electrical isolation gained by using optical coupling reduces the risk of erroneous firing of the lasers.

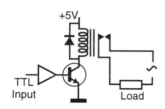

Fig. 3.20 Driving a high current device such as a relay using a transistor buffer. The relay can then be used to switch other devices such as mains powered heaters, motors or lamps.

When driving inductive loads (such as relays or solenoids) with transistors, it is necessary to place a reverse biased diode in parallel with the inductance to stop reverse emf damaging the transistor when the current is turned off.

Current drive

Although most digital circuits are protected against short circuits (*i.e.* connecting an output directly to one of the supply rails), it is generally not a good idea to try and draw too much current from a digital output for an extended period of time. If a high current device needs to be driven, then a buffer of some form should be placed between the output and the device. The simplest form of such a buffer is a single transistor. Fig. 3.20 shows how a transistor can be used to increase the current drive capabilities of a digital output. In this case, the driven device is a relay. A relay is useful in digital circuits because it allows a digital output to switch mains voltage circuits such as high power heaters or motors.

Filtering and noise

There is much written on electrical noise, both on its source and its elimination. As a reader of this book, probably the best place to start is the treatment given by Wayne in *Chemical Instrumentation*. This is an excellent introduction to the sources of noise in chemical experiments, and the methods by which its impacts can be reduced. There is little point in repeating here what is treated in greater depth by Wayne.

From the point of view of computers in chemistry, noise has a number of implications. These range from spurious pulses in counting circuits to incorrect values from ADCs. Being aware of the presence of noise is, of course, half the battle, and making sure that the circuit elements are suitable for the task being undertaken will help a great deal. In general, any signal should be filtered so that frequencies that are not of interest should be removed. This does not mean that there should be a vast array of filters to remove everything but the one frequency, more that if the signal of interest is around 1 kHz, then there is little point in having frequencies higher than, say, 10 kHz present. More importantly perhaps, the presence of interference from ac mains should be considered; if at all possible an experiment should be designed so that any oscillations are not close to the mains frequency (*i.e.* 50 Hz in Europe or 60 Hz in the U.S.) or a multiple of it (room lights actually flicker at 100 Hz, not 50 Hz).

A further consideration is the affect that noise on signals generated *by* the computer may have on other parts of the experiment. Within computers there are a wide range of frequencies present – from the 50 Hz mains to the 400

MHz CPU frequency. All modern computers are well shielded so that none of this radiation normally leaks out of the case, but when voltages and so on are generated within the computer there is always the possibility that they may have *digital noise* superimposed on them. This digital noise comes about because of the different current drawn by a logic circuit depending on the state of its inputs and outputs; since the power supplied to each chip has a finite source impedance, as the current drawn by the chip changes, so the supply voltage changes. Some of the chips, especially CPUs and the highly integrated peripheral chips found in modern PCs, can draw large currents, and with millions of logic elements all changing state at once, the changes in current drawn can be quite significant. Digital noise manifests itself in the same way as 'normal' noise, but with the difference that it has a large component at the same frequency as the CPU clock. This frequency dependence is because there are many logic gates all changing state in synchronism with the clock and so much of the interference arises at that frequency. Of course, this frequency is normally at 33, 66 or 100 MHz (in PCs), so unless the experiment is conducted at radio frequencies, it is relatively simple to filter out the noise, but it is well to be aware that these frequencies, or harmonics and sub-harmonics of it, may exist on signals derived from a computer.

Noise in general in digital circuits can be reduced by *decoupling* the supply near each chip. This decoupling consists of placing small value capacitors (~0.01 μF) across the supply voltages adjacent to every chip on the circuit board. These capacitors effectively filter out the high frequency noise on the supply voltages. This decoupling is very important near any analogue circuitry within the computer case; indeed it is often necessary to provide analogue circuits with a power supply that is isolated from that used by the digital circuits.

Within the computer case there may also be quite large high frequency fields from the CPU clock. Sensitive analogue components may pick up these fields causing problems. If it is absolutely necessary to place such components within the case, then it will be necessary to provide some form of shielding around those components. Similarly, if there are components that generate fields that may affect the operation of the computer, then they must either be taken out of the case or be shielded.

3.3 Interfacing to the PC

Chapter 2 showed in some detail the architecture of the PC. Here, details of interfacing circuitry to the PC will be presented. More specifically, details of interfacing to the ISA bus will be given. The ISA bus is usually chosen because it is the easiest to interface with, and the rate at which experimental data is collected is most often slow enough for the ISA bus to cope with. Interfacing to other buses, such as the PCI bus, for high speed applications is similar in concept, but is beyond the scope of this book; if high speed data acquisition is required, then it is probably much more satisfactory, in both quality and time consumption, to purchase a commercial interface.

The circuitry for interfacing to the ISA bus is well known and a typical example is shown in Fig. 3.21. All the connections to the computer busses are buffered so that the rest of the interface circuitry can not interfere with

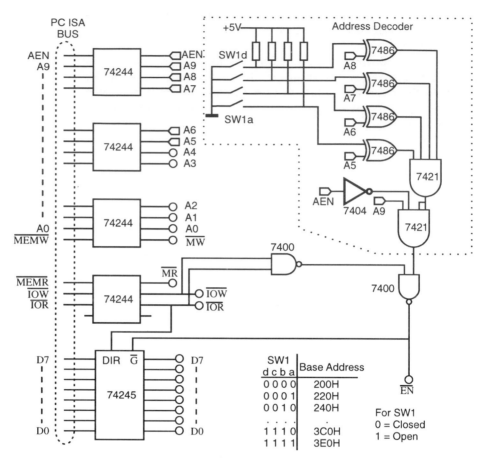

Fig. 3.21 A typical example of a circuit used to interface to the ISA bus in a PC.

the operation of the computer itself; these buffers are either uni-directional, for the address and control lines, or bi-directional for the data bus. The main feature of this circuit is the address decoder circuit at the upper right. This decoder effectively compares the address bus lines A8–A5 with a value set by switch SW1: when the pattern on those lines matches that set in the switches, all the outputs of the XOR gates will be high, and so the output of the AND gate to which they are all connected will also be high. This output is then ANDed with the A9 and AEN (address enable) lines to give a signal which will be high only when the correct address range is accessed. Since A9 has to be high, and only lines A8–A5 are compared, this circuitry will allow address in the range 200H–3E0H to be decoded in groups of 20H; *i.e.* the possible decoded base addresses are 200H, 220H, 240H... 3C0H, 3E0H. This decoded base address signal is further gated by the $\overline{\text{IOR}}$ and $\overline{\text{IOW}}$ (the I/O read and write) signals to produce the final $\overline{\text{EN}}$ (enable) signal for the rest of the circuitry.

Although in essence any address on the I/O bus may be chosen for your device(s), there are a number of address ranges set aside for 'end user' use. In this way it is less likely that you will choose an address already occupied by another device, such as the video or hard disk controller. Nevertheless, before choosing exactly which addresses you are going to use, it is wise to check that there is nothing else residing there! The address ranges reserved for your use is 300H–380H.

The bi-directional buffer on the data bus needs some explanation. This device allows signals to flow in either direction depending on the state of the 'DIR' pin, or isolates the two sides depending on the gate ('$\overline{G}$') pin. Consequently, the gate signal is derived from the $\overline{EN}$ signal (*i.e.* the data bus is isolated unless the address range is selected) and the direction is controlled by the $\overline{IOR}$ signal.

At this point, it may be worth mentioning the actual board used inside the computer. Although the physical dimensions of such a board are well defined by the PC standards, it may not always be convenient for such a board to be fabricated in a laboratory. However, there are a number of alternatives. First it is possible to buy 'PC Breadboard' cards; these are cards which are of the correct dimensions and have an edge connector for connecting to the IO slot; they also often have address decoding circuitry built on to them, but the rest of the card is blank and can be used to build your own circuits. Secondly, it is possible to buy (or fabricate) interface cards; here only a small card containing address decoding circuits is actually placed in the computer, the resulting signals are presented on a connector on the back of the computer where they can be taken to another box that contains the actual experiment interface circuits. In general there is little to choose between these two approaches, the former produces a neater end result but may present problems during prototyping and if there are many connections to the external equipment, whereas the latter is easier to construct but will result in (yet) another box! There may be overriding concerns when choosing the method of construction: there is only a limited amount of power that may be drawn from the computer's supply, hence any high power circuitry will have to be built in a box with its own power supply; similarly circuits that may be sensitive to the high frequencies present in a modern computer may be better built in an external enclosure.

As we have just seen, this particular interface board provides decoded address in 32 byte groups, others may provide 16 or 8 byte groups depending on how many address lines are decoded. The advantage of using larger groups of addresses is that if the signals are taken to an external enclosure, then it is a relatively easy job to further decode the address lines to provide smaller groups (as described later in Section 3.4). Conversely, if only small groups of address are used, then there may be insufficient addressing capacity in the external box. The decoded address ranges are then used to 'select' each interface chip as needed (*via* the 'chip select' pin), and so each chip will have a unique address range. The individual addresses, or registers, within that chip are then selected by applying the requisite number of low address lines to the chip. Take for example the 8255 PIO chip described in the previous section. The chip select pin would be connected to the output of the address decoding circuitry corresponding to the required address range (say 310H), and address lines A0 and A1 would be connected to the register select inputs; the individual registers would then appear at sequential addresses. The final signal that needs to be connected is R/$\overline{W}$: this line will enable the chip to respond correctly depending on whether you want to write information to the chip or read from it.

Read/write signals come in two different types: the combined R/$\overline{W}$ and the separate $\overline{RD}$ and $\overline{WR}$. The signals used by a particular chip depend on its manufacturer and the application for which it was developed. Translating between the two is usually a simple matter.

If there are more devices in the circuit, each would be connected to its own decoded address range, and the address lines would be connected as required. Other signals are also connected if needed: the data bus is usually required (although not if just the action of accessing an address is all that is needed), the $\overline{RD}$ and $\overline{WR}$ (or R/$\overline{W}$) lines, the RESET line (so that the chip will be in a defined state when the computer is reset), and, sometimes, the computer clock. And that is all there is to it, the rest is programming!

3.4 A worked example

The final section of this chapter will be spent examining the hardware of a real laboratory interface. The example chosen is one that encompasses many of the devices introduced so far. The actual device is used to interface a PC to an experiment in which a mass spectrometer detects radicals in a gas flow system. The details of the actual experiment are unimportant, all that it is necessary to know is that the output of the mass spectrometer consists of a series of pulses (each pulse is an ion that reaches the detector), the mass that is selected is controlled by a voltage, and the gas flows are sensed using devices whose output is a voltage linear with flow. Various other on/off functions are also required, such as switching photolysis lamps on and off, sensing the presence of cooling water flows and so on. There are thus at least four distinct functions required: counter, ADC, DAC and PIO.

The counter chosen for this device is actually a very complex chip: the Am9513 counter/timer chip. The advantage of using this particular chip is that it not only includes a number of large counters, but it also contains a number of programmable timers. The timers are important in accurately measuring the *rate* at which pulses arrive at the input, and the chip contains within it the necessary functionality to automatically count for a predetermined length of time.

The analogue input voltages are measured using four 12-bit ADCs (AD7578), whilst the output voltages are generated by four 12-bit DACs (AD7548). Finally, the PIO device is the same as the one described earlier, *i.e.* the 8255.

The simplified circuit diagram of the interface is shown in Fig. 3.22. The individual parts of the circuit have been grouped together, with each section being built on a separate card within an external interface box. The cards are connected together by a 'backplane' that carries the data, address and control busses. These busses are shown to the far right of the circuit diagram and are connected directly to the interface card in the host PC *via* a multiway cable. This arrangement means that circuits can be altered and replaced in the interface box easily without disturbing unrelated functions.

The method of fabricating the circuit boards is unimportant and depends mainly on the materials and facilities available, to say nothing of the abilities of the fabricator! The boards in this case were actually custom designed printed circuit boards, but there is no reason why they could not have been made using wire-wrap techniques or prototyping boards such as 'Veroboard'.

It is probably useful to point out some important features of the circuit. On the counter board, the first circuit that the input encounters is a *discriminator*. The effect of the discriminator is to pass through only pulses whose voltage is above a certain value, so rejecting, or discriminating, low level pulses

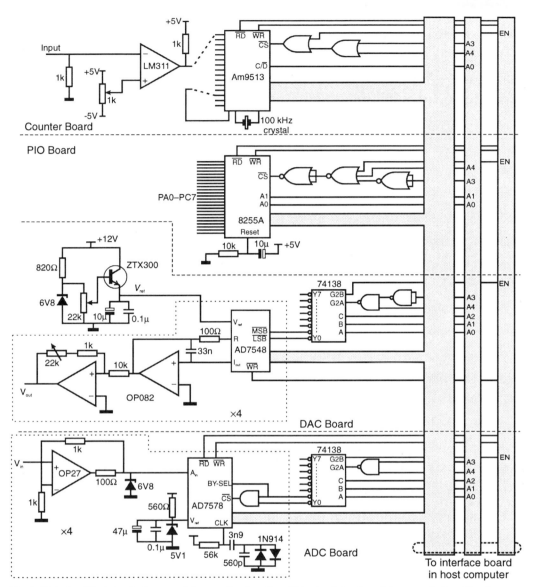

Fig. 3.22 A simplified circuit diagram of the complete interface. This circuit is adapted from one designed by the Electronics Workshop of the Physical & Theoretical Chemistry Lab., Oxford University.

which are usually just noise. The level at which the discriminator rejects pulses is set by the variable resistor. The counter chip has only two addresses (or ports) and so only the A0 line is connected to it, the A3 and A4 address lines are combined, along with the $\overline{\text{EN}}$ signal, to provide a chip select (CS) line. The crystal is part of the circuitry of the chip that provides a stable reference frequency for its timing functions.

The PIO card appears very simple. The only thing of note is the use of multiple-input gates with all the inputs tied together. In this case, the triple input NOR gates are acting as inverters, and this arrangement is purely to reduce the package count on the circuit board. If this 'trick' weren't used, it would be necessary to have another chip present which provided the inverters. The 'real' card is actually much more complicated, but only because there are LEDs on the front panel to show the state of all 24 outputs, and the circuit to drive these LEDs takes up much of the card!

For clarity, only one section of the DAC card is shown; there are actually four identical circuits providing four independent output voltages. The reference voltage circuit, shown to the left, is shared between the four DACs. This particular circuit provides an adjustable reference voltage. The two operational amplifiers convert the output current of the DAC into a voltage (*c.f.* Fig. 3.6) and then buffer it. This buffering is so that any current drawn by the external device to which the circuit is connected does not affect the actual voltage generated.

Again, for clarity, only one section of the final card, the ADC card, is shown. The input voltage is first buffered before being fed to the ADC. The feedback circuitry around this buffer determines the input voltage range. The zener diode on the output of the op-amp protects the input of the ADC from excessive voltages. The reference voltage for the ADC is derived from another zener diode, and, since this is a successive approximation converter, the required clock signal is derived from a simple oscillator circuit at the bottom of the ADC.

Obviously, each of the individual sections is located at different addresses in the memory map, as determined by the decoding of the A3 and A4 address lines. This means that each card has allocated to it eight sequential addresses. The memory map for the whole interface is shown in Table 3.2. It is assumed in this diagram that the address of the interface in the PC is set to 300H.

Table 3.2 Addresses used by interface

Address			Function
300H	768D		Data register
301H	769D	Am9513	Control register
302H–307H	770D–775D		3 Repeats of 300H–301H
308H	776D		Port A
309H	777D		Port B
30AH	778D	8255A	Port C
30BH	779D		Control
30CH–30FH	780D–784D		Repeat of 308H–30BH
310H	785D		DAC1 LSB
311H	786D	4 × AD7548	DAC1 MSB
312H–317H	787D–791D		310H–311H repeated for each DAC
318H	792D		ADC1 LSB
319H	793D	4 × AD7578	ADC1 MSB
31AH–31FH	794D–799D		318H–319H repeated for each ADC

As can be seen, some of the ports appear multiple times. This is a consequence of not decoding fully all the address lines on some of the boards. This repetition is not a problem, and usually only the first instance of each address range is used, although using the other ranges is possible.

The details of programming this interface are dealt with in the next chapter.

4 Programming the hardware

The interface hardware described in the previous chapter is, of course, useless without some form of program to control it and present the data. This chapter will describe some of the techniques used in fulfilling that rôle.

Once again though, it must be stressed that the examples and techniques presented here are not meant to be a definitive treatise on programming, rather they are a reflection of the author's own knowledge and the hardware and software available to him.

4.1 The basics

We saw in Chapter 3 how the registers of an interface device occupy specific locations in the memory map of a microprocessor. In order to access or modify the data held in those registers it is necessary to simply read from or write to those memory locations. However, the consequences of those reads or writes is dependent entirely on the specific device in question. For instance, the action of reading data from an ADC might start the next conversion, whilst a different type of ADC may require a bit in a register to be toggled to initiate the conversion sequence. The key is to thoroughly read, and understand, the data sheets and programming instructions of any device used.

The choice of programming language used depends on many factors: an idealistic solution would be to weigh up the pros and cons and choose a language most suitable to the job in hand. More often than not though, the choice is proscribed by the languages available on the computer you are using, and, as an over-riding factor, the abilities of the programmer. Nevertheless, if there is still scope for choice, there are two rough guidelines:

- interaction at the chip level is best done using machine level instructions;
- peripherals such as mass storage or display devices are best addressed using high-level languages.

Needless to say, these are rules that are just crying out loud to be broken! Many high-level languages have extensions specifically designed to make interaction at the chip level easy; similarly, there are well documented methods of performing high-level I/O functions from machine code programs. Using the 'ideal' situation of a mixture of high- and low-level languages can also be a problem, not least of which is reliably passing information from one part of the program to the other.

A further factor that the programmer should be aware of is how other programs or hardware in the computer may affect the running of their code. The main thing to be wary of is a multitasking environment. In rudimentary multitasking systems, such as Microsoft Windows, it is necessary that programs are 'well behaved' and regularly relinquish control of the machine to other programs. Not only do you have to be careful to co-operate and allow other programs some time, but you also have to be aware that other programs will *not* be as well behaved as you want. Specifically, it is very unwise in such a situation to rely on the fact that you will be able to perform a certain action within a particular time. If at all possible, especially for novice programmers, it is much easier to program in a non-multitasking environment such as DOS.

The next few sections are devoted to a brief survey of the abilities of various high-level languages with regard to interfacing, whilst the following section will introduce machine level programming. The primary platform used will once again be the IBM PC, but the techniques are equally applicable to other platforms.

4.2 BASIC

One of the easiest languages to program in is BASIC. This, coupled to the fact that there are many excellent implementations of the language available for many different computers, makes it one of the most popular languages. Many versions have specific instructions for accessing memory (both conventional and I/O) directly.

The standard BASIC commands for accessing memory directly are PEEK and POKE. PEEK is a function that takes as its argument an address, and returns the contents of the memory at that location; POKE takes two arguments, the first being an address, and the second a value to store at that address. In the PC, the analogous functions INP and OUT perform the same functions on the I/O address space.

The range of valid values for the arguments to these functions is dependent to a large extent on the specific implementation of the language. In general though the value read from or written to memory is usually 8-bit (*i.e.* in the range 0–255) although it can, occasionally, be 16-bit. The address range is much more variable. In the PC, the value is 16-bit (*i.e.* 0–65535), but for the main memory, this value does not cover all possible addresses. The reason lies in the way that memory addresses are formed and in the way that the x86 chip uses the DS register. Normally it is not necessary to worry about the value of this register, but in cases where there is a need to use a specific part of memory (if, for instance, an interface card makes data available in main memory as opposed to I/O memory), then the segment of memory accessed by the PEEK/POKE instructions can be set using the 'DEF SEG' instruction.

In x86 chips, the real address of any data in memory is always formed by shifting the DS (data segment) register left 4 bits, then adding that value to the one specified in the machine code. *e.g.* if DS=1E00H, then the instruction 'MOV AL, [200H]' would not read the contents of location 200H, but instead it would use location 1E000H + 200H (=1E200H).

There are also instructions in most implementations of BASIC to execute code that isn't part of the main BASIC program, usually some variant of the 'CALL' statement. It is important to read the manual carefully when it comes to the 'CALL' command, especially if there are parameters to pass between the different pieces of code. Each language has its own idea about how to pass variables; for instance, are the parameters passed by value or by reference, what order are the variables passed, how are arrays or strings handled, and so on. When trying these sorts of things for the first time with any combination of languages, it is always best to experiment under controlled circumstances to ensure that each part of the program is receiving the data properly.

A typical example of a short BASIC program to interact with an interface is shown in Fig. 4.1. This program uses the interface developed at the end of Chapter 3 to output a sawtooth waveform on one of the DAC outputs. The waveform is generated simply by loading the DAC register with an incrementing value; the PIO is used solely to show what is happening *via* the front-panel lights.

```
DECLARE SUB WDAC(n%)                Declare subroutine

pio% = 776
adc% = 792                          Initialise variables
dac% = 784

OUT pio% + 3, &H80                  Initialise PIO

10 FOR i% = 0 TO &HFFF              Start loop

    CALL WDAC(i%)                   Call subroutine to output value to DAC

NEXT i%                             Repeat loop
GOTO 10                             Ad infinitum
END                                 End of main program

SUB WDAC(n%)                        Start subroutine
    j% = (n% AND &HF) * &H10        Extract the low and high bytes
    k% = n% \ &H10
    OUT dac%, j%                    Output values to DAC
    OUT dac% + 1, k%
    OUT pio%, j%                    Output values to PIO
    OUT pio% + 1, k%
END SUB                            End of subroutine
```

Fig. 4.1 A simple program that interacts with an interface on the I/O bus.

The first OUT statement is just setting up the PIO, whilst the main body of the program is indicated clearly. Note the use of a variable to hold the addresses of the DAC and PIO – this is not just to make the program more understandable (which it does), but also, during development work the address of the individual components may change: making the address a variable means that you only have to change the address in one place. Note also the use of a subroutine to actually alter the value of the DAC – in this case it is not really necessary, but in a larger program, where you may be changing the value of the DAC in many places, it makes sense to only write the code once.

4.3 C

The language C was originally developed in the Unix environment – indeed the Unix operating system is largely written in C – but it has become a popular language in other environments. Because of its roots in larger, multitasking, environments, there are no specific functions in C to interface with hardware devices directly. However, since the language was designed to be used for writing operating systems, it has well documented interfaces to machine code subroutines. Consequently, most hardware interfacing in C is done using small machine code subroutines. Some implementations of C even have an *in-line* assembler – this means that the C compiler is able to assemble small sections of assembly language code written directly into the C program, and incorporate the machine code produced directly into the executable. Having said this, some implementations of C for the PC do contain extensions that allow you to access the I/O address space. For instance, Borland's TurboC uses the 'inport' and 'outport' functions, but these functions are trivial to write in assembly language and so can easily be duplicated.

4.4 Fortran

Fortran is by no means a 'trendy' language; in fact it dates from around 1956. Despite this, it is still the language of choice for many scientists. Again, there are no specific instructions for interfacing with hardware, but use can easily be made of external machine code subroutines. The main advantage of Fortran is that it is relatively easy to perform complex calculations, such as Fourier transforms, on the data obtained from an experiment. One problem with Fortran is that the passing of parameters between subroutines is not as simple as might be hoped, but careful study of the manual, and the example programs supplied with it, should make things clear.

4.5 Assembly language

The code that the CPU executes, the *machine code*, is stored as a series of binary numbers in memory. Each binary number has encoded in it both the operation for the CPU to undertake and the data, or a pointer to the data, on which to perform that operation. For instance, in the 8086 processor, the four bytes '88C3' means 'move the contents of register AL to register BL'. Although the structure of the machine code instructions is logical, it is not memorable, and the encoding is designed solely to be easy for the CPU to interpret. In order to make things easy for mere humans to deal with assembly language is used. Assembly language is a set of mnemonics for the CPU operations; these mnemonics are translated, with an *assembler*, into machine code instructions. The assembly language instruction for the above machine code would be 'MOV BL, AL'. Each type of CPU has its own set of assembly language instructions: assembly language written for an 8086 processor would not be compatible with that for a 6800 processor. As usual, the 8086 processor, *i.e.* the processor used in the IBM PC, will subsequently be used as an example.

It is certainly not the intention to provide a detailed course on programming in assembly language here, there are plenty of excellent texts that can be used. However, an introduction to some basic principles is in order.

Addressing modes

It is a basic function of a CPU that it can perform functions on both its internal memory (or registers) and on external storage locations. In assembly language, operations on registers are specified by using the mnemonic for the register, e.g. 'MOV AX, 200H' means move the value 200H into the register AX, whereas operations involving memory are specified using a variable name corresponding to a location, e.g. 'MOV AX, TOTAL' means move the value in memory location pointed to by 'TOTAL' into the register AX. These different ways of accessing memory locations are called addressing modes: the two encountered so far are called *immediate* and *direct* addressing – where only registers are involved, such as shown above, it is called *register* addressing. These modes are all that are needed for simple programs, but where more complex functions are necessary, more complex addressing modes are employed.

Indirect addressing is where the location of the address of the operator is given to the processor: in the 8086 series of processors only the BX, BP, SI or DI registers may hold that address, other processors can use a memory location. As an example take the instruction 'MOV AL, [BX]', this means move the value stored in the memory location whose address is held in the register BX into AL. A more complex form of indirect addressing is *indexed* addressing. Here the effective address of the operator is made up of the sum of a number of different parts: *e.g.* the BX register plus the SI (or Start Index) register plus a constant.

These more complex addressing modes are useful when writing machine code subroutines for use with higher level languages. Often it is necessary to pass variables to and from the subroutines, and it is possible that these variables will be passed as an address that points to the variable, rather than the value of that variable – this is especially true if arrays or strings are passed. Consequently, in order to access these passed variables, it is necessary to use some form of indirect or, in the case of arrays, indexed addressing.

These indexed addressing modes show the rudiments of array operations: the 'constant', or BX, may hold the start address of an array, with the SI register holding the index of the element within the array.

Stacks

We have already met stacks in Section 2.1, but it is worthwhile revisiting them here in a programming context. As mentioned in that section, stacks are important in the operation of a processor, as they are the mechanism by which the state of the processor can be 'remembered'. Whenever program flow is transferred to a subroutine using the 'CALL' command, the processor puts the address of the next instruction to be executed after the 'CALL' onto the stack. When the subroutine executes the 'RET' statement to return to the calling code, the processor retrieves the address put onto the stack and continues execution. These two processes of placing and retrieving values from the stack are called *pushing* and *popping*.

Subroutine calls are not the only use of the stack. There are two instructions called, unsurprisingly, 'PUSH' and 'POP', that allow the programmer to push and pop registers to and from the stack. The stack is thus a convenient method of temporarily storing registers. The code segment in Fig. 4.2 shows an example of how a register may be preserved using PUSH and POP. The 'IN' instruction reads a data value from an I/O port into the AX (or AL) register, the address of that register can either be directly specified as a byte in the instruction (if the port number is less than 255), or as an address in the DX register. If it is necessary that the DX register be preserved, then it can be pushed onto the stack, and popped after the IN command.

Processors other than the 8086 series may have different stack commands, for instance there may be more than one stack, or commands to push combinations of registers onto the stack may exist. The 'CALL' command (or its equivalent) may push more than just the program counter onto the stack,

```
PUSH    DX
MOV     DX,0300H
IN      AX,DX
POP     DX
```

Fig. 4.2 Example of using PUSH and POP to preserve the value of a register.

or there may be different 'CALL' commands that preserve different groups of registers. The important thing is to be familiar with the assembly language of the specific processor you are using.

One further important feature of stacks is that they are often the mechanism by which variables are passed between subroutines in high level languages. Take for example a subroutine in, say, MS Fortran, which is called with the code 'OUTDATA(P,A)' where P is the port data, is to be written to, and A is the data which is to be written, both of which are declared as 'INTEGER * 2', or 16-bit integers. This subroutine, as its name suggests, writes the value of A to port P. When the subroutine is called, the value of 'P' is first placed on the stack, followed by A, and then the code is called. The first thing the routine needs to do is retrieve these values without disturbing the stack (as it contains the location to return to needed by the RET statement). The easiest way to achieve this is to copy data directly from the stack; with the 8086 CPU, this is achieved with the BP register – this register is similar to the BS register, but uses the SS segment rather than the DS segment as its base. The structure of the stack after the subroutine call is show in Fig. 4.3; it can clearly be seen that A is located at [SP+2] and that P is at [SP+4]. The subroutine is consequently relatively easy to construct and is shown below.

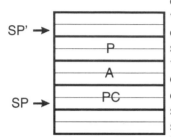

Fig. 4.3. The state of the stack after a subroutine call in MS Fortran. SP' is the position of the stack pointer before the call, and SP is the position after.

```
MOV    BP, SP        ; Copy the stack pointer to BP
MOV    AX, [BP+2]    ; Retrieve A into AX
MOV    DX, [BP+4]    ; Retrieve P into DX
OUT    AX, DX        ; Write to I/O port
RET    4             ; Return to calling code
```

Fig. 4.4 Example of subroutine to show how to retrieve data that has been placed on the stack by the calling program

One thing to note is the 'RET' statement. Normally, this statement just pops the top two bytes from the stack and uses this as a return address. However, when parameters have been passed on the stack they must be discarded after the routine has returned. This could be done by the calling code, but in the 8086 you can specify in the 'RET' command how many extra bytes should be popped from the stack in order to tidy things up and keep the stack consistent.

Finally, a word of warning concerning stacks. It is essential that any subroutine should leave the stack in an identical state to when it found it. Otherwise, if you have too many or too few 'PUSH's or 'POP's, then a 'RET' command may use the wrong return address and mayhem is bound to ensue!

Flags and jumps

The ability to 'make decisions' is one of the basic features of a computer program. Although programs that follow the same, unvarying, sequence of instructions can be useful, they are not very versatile. The decision making in a program comes from the ability to test a condition and change the program flow depending on whether that condition is true or false. In machine code, the same is true, but the conditions able to be tested and the program 'constructs' available are much more limited. Generally, the program flow in machine code is changed by jumping to a new location; that jump may be unconditional, or it may depend on certain conditions. Those conditions are usually determined by the flags.

As explained in Section 2.1, the flags are a series of one-bit storage locations grouped together into the flag register. The values of the flags are affected by the operations of the CPU. For instance, the zero ('Z') flag is set to '1' whenever the result of certain operations is zero. In the 8086, there are nine flags (see Fig. 4.5), some of which are set by the results of instructions, others are set by the instruction itself and affect how the processor performs some operations.

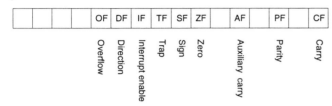

Fig. 4.5 The flag register in the 8086

In general there are usually jump instructions to cover all the possible conditions of the flags, along with some instructions that act on combinations of flags. The 8086 has 18 conditional jumps (and an unconditional one), as shown in Table 4.1; other CPUs may have more, others may have fewer.

Table 4.1 The conditional jumps in the 8086

Instruction	Condition	Interpretation (Jump if ...)
JE or JZ	ZF = 1	"equal" or "zero"
JL or JNGE	(SF xor OF) = 1	"less" or "not greater or equal"
JLE or JNG	((SF xor OF) or ZF) = 1	"less or equal" or "not greater"
JB or JNAE or JC	CF = 1	"below" or "not above or equal"
JBE or JNA	(CF or ZF) = 1	"below or equal" or "not above"
JP or JPE	PF = 1	"parity" or "parity even"
JO	OF = 1	"overflow"
JS	SF = 1	"sign"
JNE or JNZ	ZF = 0	"not equal" or "not zero"
JNL or JGE	(SF xor OF) = 0	"not less" or "greater or equal"
JNLE or JG	((SF xor OF) or ZF) = 0	"not less or equal" or "greater"
JNB or JAE or JNC	CF = 0	"not below" or "above or equal"
JNBE or JA	(CF or ZF) = 0	"not below or equal" or "above"
JNP or JPO	PF = 0	"not parity" or "parity odd"
JNO	OF = 0	"not overflow"
JNS	SF = 0	"not sign"

As an example of the use of flags and jumps, the machine code for reading the ADC in the example interface will be examined. The code is shown in Fig. 4.6. A conversion is initiated by writing to the ADC result register, and the top bit of that register remains set until the conversion is finished. It is thus necessary to wait until that bit is cleared before reading the result. The result itself is given in two consecutive locations, the high byte being in the first. As we can see, after the conversion is started, the ADC is read and an AND operation with 080H is performed on the result (the TEST instruction performs an AND, but does not change the contents of the register – the flags are set according to the results of the operation though). This operation

Fig. 4.6 The code segment for
reading an ADC showing the use
of a conditional jump

```
            MOV     DX, ADC          ; put ADC in DX
            MOV     AL, 0            ; initialise AL
            OUT     DX, AL           ; start conversion
    LOOP    IN      DX, AL           ; get ADC register
            TEST    AL, 080H         ; test high bit
            JNZ     LOOP             ; if set, try again
            MOV     AH, AL           ; move to high byte
            INC     DX               ; point to next register
            IN      DX, AL           ; read low byte
                                     ; result now in AX
```

Fig. 4.6 The code segment for reading an ADC showing the use of a conditional jump

effectively *masks* the single bit we are interested in – the result of the AND operation will be zero if that bit is '0', and non-zero if that bit is '1', and the Z flag will be set accordingly. Thus, since we are waiting for that bit to become '0', we use the 'JNZ', or 'jump if not zero', instruction to cause a jump to the label 'LOOP' to occur when the Z flag is not set. Finally, when the conversion is finished, the 'AL' register already contains the high byte of the result, so it is moved to the 'AH' register, the port pointer 'DX' is incremented to point to the next port, and the low byte of the result is read into 'AL' – the final converted value is thus in 'AX'.

In the 8086 series, as in many types of processor, short jumps can be made either forwards or backwards; short in the case of the 8086 means ±127 bytes. Longer jumps can be made, but they can not be conditional, *i.e.* only the 'JMP' command can be used. This may seem to be a considerable restriction, but in practice it will not be much of a problem: if you are only writing interface subroutines, then you will rarely write more than 100 bytes of code, but if you do need a long conditional jump, then just use a 'JMP' command with a conditional jump around it (see Fig. 4.7).

```
        JNE     SHORT
        JMP     LONG
SHORT ....
```

Fig. 4.7 Code example to show programming a long conditional jump

4.6 Example project

It is difficult, in a primer as this, to give a complete example of an interface project. Nevertheless, we have already been introduced to the hardware side of the interface, it is now time to examine the software. The task that the program has been developed for is somewhat contrived and will mainly be used to show the techniques involved. The task is: on a given signal (*i.e.* an external switch closing), a time delay is initiated, at the end of the delay a voltage is sampled, and that voltage is then output on a DAC. The whole system is thus a sort of sample-and-hold and is designed to use all parts of the interface. Fig. 4.8 shows a flow chart of the system. The final code is programmed in QBASIC (the Microsoft version of BASIC that is bundled with DOS).

The programming of the Am9513 is too complex to describe in detail here. For more in-depth explanations, the data sheet and programming manual for the device should be studied.

The first part of the program is the initialisation. The ADC and DAC do not need any initialisation and the 8255 PIO needs only the control register setting, the Am9513 however needs extensive initialisation and is set-up so that one section generates 1kHz whilst another is set to be a count-down timer. This counter initialisation is not shown here.

The next section is code to wait for the 'start' signal: this signal is a high-level voltage on bit 1 of port A of the PIO, and could be either a switch closing or an output from another device. The subroutine merely returns the value of the required bit, with the 'decision' being taken in the main program. Once bit A1 goes high, the program calls the routine to start the delay timer. The delay end is not sensed at this point, as it is known that it is some time away, and so control is returned to the main program and a brief message is printed.

Once the delay period has finished, the subroutine to perform the analogue to digital conversion is called. This routine first initiates a conversion, then waits for it to finish and finally reads the result and returns to the main program. Once the value has been passed back, it is used to set the output voltage using the next routine. Finally the value is printed out and the program loops back and waits for another event.

This program is written purely in BASIC; however it need not be so, nor is it really necessary in such a short program to split things into subroutines and procedures. However the rationale behind splitting up the program in such a way is that if necessary, the sections that interact directly with the hardware can be written in machine code and directly replace the BASIC ones. If this was done, then one of the reasons for splitting the delay routines becomes apparent: in a multitasking system, when the processor is executing machine code subroutines that don't involve a system call, no other process is given time to execute. If the machine code subroutine often returns to the controlling program, then other programs get a chance to execute. A similar reasoning is behind making the decision on whether the PIO bit is set in the main program rather than in the subroutine: *i.e.* so that the processor isn't trapped in the subroutine until something happens externally.

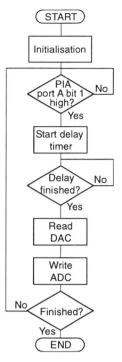

Fig. 4.8 Flowchart of example project

```
DECLARE SUB init ()
DECLARE FUNCTION pioAb1% ()
DECLARE SUB dlystart ()
DECLARE FUNCTION dlyend% ()
DECLARE FUNCTION readADC% ()
DECLARE SUB wrDAC (value%)

COMMON SHARED cntd%, cntc%, pio%, dac%, adc%

cntd% = 768: cntc% = 769: pio% = 776: dac% = 784: adc% = 792

CALL init                       ' Initialise hardware
DO UNTIL INKEY$ <> ""

        DO UNTIL pab1% = 1
                pab1% = pioAb1%
        LOOP                    ' Wait until Port A, Bit 1 is 1

        CALL dlystart          ' Start delay timer
        PRINT "Delay started"
        d% = 0
        DO UNTIL d% = 1
                d% = dlyend%
        LOOP                   ' Wait for delay to end
        PRINT "Delay finished"

        adcin% = readADC%      ' Get ADC value
        PRINT adcin%
        wrDAC (adcin%)         ' Set DAC

LOOP                           ' Repeat until a key is pressed
END
```

Fig. 4.9 Example interface program

```
SUB init
OUT pio% + 3, &H90      ' Set PIO for Port A input

CALL init_count
END SUB

FUNCTION pioAb1%
portA% = INP(pio%)          ' Read Port A of PIO
pioAb1% = portA% AND &H1  ' Mask of bit 1 and return it
END FUNCTION

FUNCTION readADC%

OUT adc%, 0                      ' Start ADC

b% = INP(adc%)
DO WHILE ((b% AND &H80) = 0)
        b% = INP(adc%)
LOOP                          ' Wait for ADC to finish
a% = INP(adc% + 1)           ' Get low byte
b% = b% AND &HF
readADC% = a% + &H100 * b%    ' Return 12-bit number
END FUNCTION

SUB wrDAC (value%)
a% = value% AND &HFF                ' Extract low byte
b% = (value% / &H100) AND &HFF   ' and high byte
OUT dac%, a%                      ' Output low byte to DAC
OUT dac% + 1, b%                   ' Output high byte to DAC
END SUB
```

Fig. 4.9 (cont.) Example interface program

5   Software for the laboratory

The previous chapter introduced some of the programming techniques used in creating the software used in instrumentation. In this section, we will look more at what the software does.

5.1 Background

It will come as a relief to most people that it is not necessary to write all the instrumentation software yourself; indeed it is easily possible to produce a working, useful system to control an instrument from a computer using off-the-shelf hardware and software. Further, most modern instruments were designed with computer control in mind, and many are essentially useless without that computer control. Take as an example the development of UV-Vis or infra-red spectrometers. Ancient instruments (*i.e.* pre 1970's) were generally electromechanical in design and completely stand-alone. Now, similar instruments have either inputs and outputs to allow interaction with external devices, or are completely dependent on computer control.

This computer control usually takes the form of a general-purpose computer such as a PC. There are two ways that these sorts of instruments can be configured for computer control: the spectrometer may be completely 'dumb' and the attached PC controls every minute detail, or conversely the spectrometer may have a degree of 'intelligence' and the PC only gives it general instructions. It is important to understand the implications of these two scenarios. If the computer is controlling even the basic functions of the instrument, then often things happen in a time-critical manner, and the computer must be available to respond when necessary. The consequence of this is that the computer will be dedicated to this one task, and can not be used for other things, such as data manipulation, while it is scanning a spectrum. If the spectrometer is intelligent, then the computer may be able to give it general commands, like "scan from 300 nm to 400 nm at 1 nm s^{-1}", and then resume other tasks while it is waiting for the scan to finish.

This arrangement of one computer giving instructions to another is often called a *master/slave* set-up. The master computer, the PC, decides on the general strategy, while the slave performs the actual tasks. The slave computer may often be as powerful, if not more powerful, than the master, especially when complex actions such as Fourier transforms are required, but it is still only a slave to the master's desires. In some instruments, the slave computer may not be physically located in the instrument: the slave may be an interface card in the PC, the advantage of this being that the transfer of large amounts of data between the master and slave is much faster.

Most modern instruments, be they spectrometers or pH meters, have some form of intelligence built into them and many are capable of working in a master/slave environment, even if they have enough built-in intelligence to be stand-alone machines. Further, those that are expected to be slave machines will often have the basic abilities to act on their own. Some large instruments take this master/slave principle a stage further, with many slave processors being controlled from one master: for instance, in a modern NMR spectrometer, there will be one controlling computer, but there may be slave processors that control the magnetic fields, the cryogenic system, the detectors, the pulse generation, the Fourier transform calculations and so on.

One problem with all this computerisation of instrumentation is: how do you get the data from an instrument to a computer, and once it is there, what do you do with it! The rest of this chapter will deal with these two thorny issues.

5.2 Linking computer and instrument

There are very few standard ways of connecting a computer to an instrument; often this is because some of the processing is performed on an interface board inside the computer, and the link between the board and the device is very specialised. Since it is very unwise to interfere with such specialised situations, we will not discuss them further. General-purpose interconnection techniques fall into two main categories: serial or parallel.

Serial interfaces

A serial connection is one in which each bit of the data is sent individually down a single line, a '1' being represented by a high voltage on the line, and a '0' by a low voltage. The specific voltages involved are well defined, but vary depending on which standard is being used (see below). It is possible for serial connections to consist of just two wires (the data connection and a ground for reference), but this means that data can only be transmitted in one direction at a time, so called *simplex* connections. Most serial connections these days actually consist minimally of three wires, one each for data travel in each direction, so that transmission can occur in both directions at once, and a ground. This arrangement is called a *duplex* connection. There are often many more wires in a serial connection, and these are used for data flow control – so that, for instance, the receiving end can indicate to the transmitting side that it is not ready to receive data yet and so on. These *handshaking* signals, although not absolutely necessary, enable much more efficient error free transfer of data and allow transmissions at much higher speeds. Serial connections most often use 25-pin or 9-pin 'D'-connectors, and the pin-outs of these connectors are shown in Table 5.1. These

Table 5.1 Pin-out of RS232 serial connectors

Pin Number		Legend	Description
25-pin D	9-pin D		
2	3	XMT	Transmitted data
3	2	RCV	Received data
4	7	RTS	Request to send
5	8	CTS	Clear to send
6	6	DSR	Data set ready
7	5	—	Ground
8	1	DCD	Data carrier detect
20	4	DTR	Data terminal ready
22	9	RI	Ring indicator

connections and connectors were developed in the days when terminals were connected to computers through serial lines, consequently many of the connections are redundant now. Other types of connectors are sometimes also employed on computers, but if this is the case, then adapters to a 'D'-connector is often supplied.

A serial connection between computer and instrument usually uses the built in serial ports on the computer. Most often this type of interface is used in situations where the rate of data acquisition is low, or there is a small amount of data produced. This is because serial ports are inherently slow and the rate of data transfer is not very high. Take for example a serial port running at 9600 baud; each bit takes 1/9600 s (or approximately 0.1 ms) to transmit, so an 8-bit character (with overheads) takes about 1 ms; so if each data point consists of two numbers occupying in total 10 characters, a maximum acquisition rate of 100 points/s is only possible. This may seem like a reasonable acquisition rate, but basic rates of 10 times this are often necessary. Of course, if the instrument has some form of built-in intelligence, data can be stored in the instrument and some form of pre-processing performed before transfer to the computer. Nevertheless, serial transfer can still be relatively slow: a 1000 point spectrum, for instance, will take over 10 s to transfer to the computer.

Serial connections are used though and they do have distinct advantages. First, and probably most importantly, they are simple and relatively easy to program: at a low level, sending characters along a serial line is as simple as writing a byte to an I/O port, with reading characters being just as simple; at a higher level, there are well documented programming interfaces to serial ports. Secondly, the electrical characteristics of a serial line are well understood and the distance between the computer and the instrument can be quite large – standard RS232C lines (like those used in the PC) can be used over a distance of about 10 m, longer distances can be achieved using current-loop techniques.

Baud rates higher than 9600 are, of course, possible, and higher rates will enable data to be transferred much more quickly. But if higher speeds are used, then the maximum length of cable will be reduced, and it is imperative to use some form of handshaking.

Parallel connections

As the name suggests, multiple bits of data are transferred at a time when parallel connections are used. The most often encountered type of connection is 8-bit parallel, meaning that 8-bits are transferred at a time.

All PCs are equipped with at least one parallel connection – the printer port – and some interfaces do indeed use it for connecting to instruments. But there are two disadvantages with this. First, often the port is uni-directional, *i.e.* the port can only transmit data, not receive it, and secondly, it is often used for a printer and so will not be available.

The most common type of parallel connection used to connect instruments and computers together is the *IEEE-488* or *GP-IB* bus. This system is not just a straightforward parallel connection between two devices, it is a multi-device bus arrangement. Each device on the bus has an address and any two devices may talk to each other at any one time. When two devices talk, one is

Baud: a unit of measure of data flow, equivalent to bits s^{-1}.

Data transmission standards: a number of different standards exist, with the most common being RS232C. This defines a '1' as being +15V, and a '0' as –15V. Others are RS422, which uses ±5V, and current loop, which uses a current of 40 mA to represent a '1' and 5 mA for a '0'. Over long distances, the resistance of the connecting wire can attenuate the voltages used in RS232 & RS422 lines and so possibly cause errors to occur. Current loops do not suffer from this problem.

a master, and the other a slave; the master controls the interaction and initiates the conversation with the slave.

The only problem with GP-IB systems is that they are sometimes more complicated to program than straightforward serial interfaces, but the advantages of higher speed and the possibility of controlling more than one device are usually dominant.

Which interface to use?

This is not an easy question to answer, serial interfaces are simpler and more electrically robust, but parallel systems are much faster and more versatile. In the end though, choice may not be possible. Often an instrument comes with only one type of interface and you will be forced to use that. The skill is not really in choosing which interface to use, rather picking an instrument in the first place that offers the facilities you need.

5.2 Data processing

So once you have the data in your computer, what do you do with it? Well that depends on what your data represent! However, the first hurdle is to get the data into some form that is understandable by the data manipulation programs.

Data formats

There are very few 'standard' data formats in use, most instruments produce data in a format that seems the most sensible to the designers of the instrument, but is totally incompatible to all others. This makes the design of data processing programs very difficult, and the approach used by most programmers is to provide translators for a number of different data formats into their own, proprietary, format.

The first thing to determine is how the data are represented: *i.e.* is it binary or is it text? The difference between binary and text is often a cause of confusion: the number 203 can either be represented by the 8-bit binary number '11001011' or the ASCII digits '2', '0' and '3'. In general, binary data is used where very low-level data is being transferred, such as the direct output from an ADC, and ASCII is used when there has been some form of processing already applied to the data. The main advantage of binary over textual forms of data is that it is much more compact: the number 203 can be represented in 8 bits in binary and 24 bits in ASCII. However, binary data usually has a much higher degree of structure and is much more inflexible. This structure is necessary because binary data can take any possible combination of bits and so there is no room for markers to signify the start and end of data or, say, a change in scale. In general, most data formats are some form of ASCII; it is much easier to introduce a flexible structure, markers signifying 'events' can be inserted, and, possibly most importantly, it can be read by humans. Indeed binary formats are so relatively rare, and so specialised, that they will not be discussed further here.

The most basic data format is a simple stream of numbers representing some phenomenon such as a spectrum. The data stream will usually consist of one number per line, and the format of that number will depend on the actual value of the data (it may be integer, real or in 'scientific' notation).

ASCII – American Standard Code for Information Interchange – the standard way of representing characters (numerals, letters, symbols) in a computer. Each character has a unique number based on a 7-bit binary word. For instance the character 'P' is represented by 1010000 in binary, or 80 in decimal. In 8-bit computers, the extra bit is either used as a parity bit, or for the extended ASCII character set.

Obviously, it is necessary to know beforehand what these numbers represent, since there is no way of determining from the data stream what, in the case of the spectrum, the wavelength for each data point is.

The next refinement is to insert into the data stream, at a predefined point, some information that will indicate what the rest of the data refer to. Most often this is done at the start of the data, and may consist of the number of data points to expect and other pertinent data like the wavelength range and increment in our spectrum example. Again though, this header data is usually in a fixed format and so only of use to one application.

Data are often best represented in groups: these groups may be anything from a single (x,y) data point to a series of measurements all taken at one time. These groups, or *records*, are often represented in data files as a single line of data, with each value being separated from the others by a comma thus giving rise to the name of this type of data file: *comma separated value*, or *CSV*, files. Most data analysis programs are able to read and write CSV data, either as simple (x,y) data or as more complex records. Consequently it is probably the most commonly used general purpose data format and is a good choice to use if you are writing a program which produces data that is to be read by another program.

Beyond these simple formats lies the minefield of proprietary formats – all of which seemed a very good idea at the time and whose authors were convinced would solve the world's problems. Invariably these formats are only used by a narrow range of programs in one particular field. The tragedy for scientists is that once the data have been saved in that proprietary format, it is generally unfathomable by any other program. Unless the data are exported from that program into some suitable general-purpose format, it will be almost impossible to read the data into another program. It is not unknown for valuable data to be lost because some research student has saved the experimental results in an unknown format!

Having said that most data formats are proprietary, there has recently been a move to design a general-purpose technical data format to aid in the exchange of information. This format is called *Hierarchical Data Format*, or *HDF*. One of the problems of a general-purpose format is that each experimenter has their own idea about what a data format is: some work with simple (x,y) points, whereas others may need 3D arrays, and others might use images; the individual data points may be integer, real or imaginary. Each of these formats, and more, need to be accommodated in any general-purpose system. HDF uses a system similar to a filesystem on a computer: the individual sets of data are contained within a hierarchical structure, with each set having attributes that define its contents.

Keeping track of the data

It is surprising how many experimenters treat the data they have just laboured over acquiring with little respect. Often they spend hours searching through files on hard or floppy disks looking for the raw data from an experiment they performed "just last month".

The problem is mostly one of housekeeping, but it is nevertheless important to consider. The first and most important thing is to give your data files meaningful, consistent, names. There is no point in using names such as "data9.dat" for your files: they convey no meaning and it would be necessary to actually look at the data to find out what the file contained. There are many different naming schemes, the specific one chosen probably depends on the limitations of the computer system you are using – the old "8.3" filename system used in older versions of DOS did not really encourage good naming practices! One way is to encode the date into the filename in some way as this is often the way in which results are organised: for instance you may use names such as "2001-10-17-000.dat" where the last three digits indicate the file number for that day. The use of the date in "year-month-day" format is useful, as it will mean that the files, when listed, will be sorted in date order. Another way may be to encode the experiment type into the filename: for instance experiments on ozone photolysis at 200 K may be stored in files called "o3phot-200K-000". If this method is chosen, then it is useful to store the date inside the file in some way, as the date the experiments were performed is sometimes the only way of accurately tracing the exact experimental conditions. Note that it is not sufficient to use the computer's in-built date stamping of files as a record of the date the files were created – the date on the computer may be wrong, and the dates are often not replicated when the files are copied.

The second useful technique to use when storing data is to organise the files into directories or folders. This may seem obvious to many people, but all too often you come across directories that contain a couple of thousand data files representing somebody's work over the last few years. All modern operating systems (and most ancient ones as well) offer the ability to create a hierarchical filesystem structure that can be used to create folders within folders (and so on) so that your data can be held in a logical manner. The folder structure may be based on the date (*i.e.* there is a folder for each month and subfolders for each day in the month) or it could be based on the type of experiment being performed (*i.e.* a folder for a particular compound being studied, and then subfolders for the type of experiment performed on the compound and so on). It doesn't really matter what the structure is so long as it is logical and consistent.

It is also good practice to ensure that whenever results are presented or analysed, that the name (and folder) of the original raw data file is recorded. In this way it is always possible to go back to the original data to perform any reanalysis. Further, if the original filename is constructed in a logical, consistent, manner, it will contain information on what the graph or whatever refers to.

Finally it is wise to remember that your data are precious: don't loose them. Always make sure that backups of the data are made at regular intervals and that the backups are readable. There are many ways of maintaining backups, but the simplest way is to copy the data to a floppy disk. If the data are particularly valuable or time-consuming to acquire, then make multiple backups. When making backups, ensure that any folder structure is retained: it is pointless making a backup, if that backup overwrites some previously saved data of the same filename.

Analysing the data

Before the advent of computers all data analysis was performed by hand. The technique involved getting a large piece of paper and drawing many columns on it. The raw data was entered on the left-hand side, and subsequent columns contained intermediate results calculated using either log tables or a calculator until the required final result was arrived at. The most important tools were a large piece of scrap paper and an eraser. Most of the experimenter's time was spent trying to find out where, in all these calculations, a mistake had been made!

The advent of computers meant that these repetitive calculations could be performed in the knowledge that no mistakes would be made – or that the same mistake would be made in every calculation. There are two methods of doing such calculations: a program written specifically for the purpose, or a general-purpose program customised to your needs. The most common type of general-purpose program used is a spreadsheet, and these will be examined below.

When writing a custom program for the analysis of data there are a number of things that must be borne in mind. First, even if the program is to be used only by the programmer, it must be fully commented: if a modification needs to made to the program, it can be almost impossible to try and work out what a particular constant was or where it was derived from! Secondly, check the output: it is a common failing that people all too readily accept the output from a computer program as being "right"; unfortunately, it is only as correct as the original programming. It is advisable to check the output from the program by inputing data that has been previously calculated by hand and is known to be correct; a further good test is for a second person to calculate input data backwards assuming an answer, and then checking that the analysis program does indeed come up with the correct answer. Finally, make sure that any printout from the program contains a record of what the data refer to: a sheet of numbers means nothing a couple of months later!

Spreadsheets

A spreadsheet can be likened to the Dickensian ledgers that the likes of Bob Cratchett slaved over. Numbers are entered into rows and columns, the results of calculations on these numbers are entered in to other rows and columns and so on. If one number is changed, then that change must be reflected in all the numbers that depend on it. A spreadsheet performs all these dependent calculations automatically when any entry in a *cell* is changed. Each cell can be either a value or a formula specifying how the value of the cell can be calculated.

A full description of spreadsheets is well beyond the scope of this book, but it is obvious how they can be used to help an experimenter perform repetitive calculations. A small example of a spreadsheet is shown in Fig. 5.1 (the actual data are fictitious and have been generated solely for this purpose).

The calculation in Fig. 5.1 is to measure the concentration of compound X in substance Y; compound X is extracted from Y by a solvent twice (solutions '1' and '2'). The concentration of X in the solutions is then measured. The calculation thus consists of

$$[X]_Y = \frac{[X]_{S1} \cdot V_1 + [X]_{S2} \cdot V_2}{W_Y} \qquad (5.1)$$

In the spreadsheet, the weight of substance Y (W_Y) is in column B, the measured concentrations of X in solution ($[X]_{S1}$, $[X]_{S2}$) are in columns C and F and the volumes of solutions 1 and 2 are in cells C3 and C4. The amount of X in each solution is calculated (by $[X]_S.V$) in columns D and G by multiplying the value in columns C and F by cells C3 and C4 respectively. The intermediate solid concentrations are derived in columns E and H by dividing the values calculated in D and G by the weight of the sample in column B. The total amount of compound X, derived from the sum of columns D and G is in column J, with the final concentration, calculated by dividing the values in J by the corresponding value in B, being shown in column K.

	A	B	C	D	E	F	G	H	I	J	K
1	Solvent extraction of compound X from substance Y										
2											
3	Volume of solution 1		15.0	cm³							
4	Volume of solution 2		20.0	cm³							
5											
6		Sample									
7	Expt	weight			Solution 1			Solution 2		Total	CompoundX
8		(g)	(mmol/cm³)	(mmol)	(mmol/g)	(mmol/cm³)	(mmol)	(mmol/g)		(mmol)	(mmol/g)
9	1	3.2568	27.35	410.25	125.97	15.35	307.00	94.26		717.25	220.23
10	2	3.3146	10.55	158.25	47.74	28.23	564.60	170.34		722.85	218.08
11	3	2.9899	15.62	234.30	78.36	21.95	439.00	146.83		673.30	225.19
12	4	3.0826	11.35	170.25	55.23	26.96	539.20	174.92		709.45	230.15
13	5	3.5311	30.50	457.50	129.56	12.26	245.20	69.44		702.70	199.00
14	6	3.1236	6.05	90.75	29.05	29.72	594.40	190.29		685.15	219.35
15	7	2.5998	35.08	526.20	202.40	0.74	14.80	5.69		541.00	208.09
16	8	3.4006	18.69	280.35	82.44	22.57	451.40	132.74		731.75	215.18
17	9	2.8586	8.93	133.95	46.86	24.72	494.40	172.95		628.35	219.81
18	10	3.0015	22.60	339.00	112.94	16.22	324.40	108.08		663.40	221.02
19										Average	217.61
20										Standard Deviation	8.27
21											
22			C22*C3	D22/B22		F22*C4	G22/B22			D22+G22	J22/B22
23											
24											

Fig. 5.1 Example demonstrating the use of a spreadsheet

The formulae used to perform these calculations are shown in row 22. The '$' prefix to a cell reference means that that reference is fixed (*i.e.* it doesn't change down the column, whereas the others will change); fixed cells are used here to point to the volumes in cells C3 and C4.

Finally, the average and standard deviation at the foot of column K are calculated using built-in functions. Most spreadsheets contain such functions, and in MS Excel (the one used here), the cells contain the formulae AVERAGE(K9:K18) and STDEVP(K9:K18).

The advantage of using spreadsheets over 'pen and paper' is that once a basic spreadsheet has been constructed, all that is necessary to repeat the calculation in a similar experiment is to change the raw data – in this case the data in columns B, C and F.

Plotting results

Almost all data needs to be plotted at one point or another. Experimenters are, of course, still at liberty to use graph paper and pencil, and many still do. However, it seems sensible that as most data are now either collected by, or entered into, a computer, that we use its facilities for plotting the data.

Probably the easiest way to plot data on a personal computer, be it a PC or a Mac, is to use a spreadsheet. All modern spreadsheets have some form of plotting functions built into them. However, it must be remembered that the spreadsheet was originally designed as an office, not a scientific, tool, and as such the range of scientific plotting facilities is usually very limited. For instance, probably the most popular spreadsheet package in use, Microsoft Excel, has only 1 out of 14 graph types that is suitable for plotting (x,y) data!

There are, of course, other methods of plotting data. Most LIMS packages (see below) incorporate some form of graphical capability and these systems may be the most convenient integrated way of analysing data. Dedicated plotting packages exist as well. For the PC, packages such as Origin from Microcal provide very accomplished plotting facilities along with spreadsheet style data manipulation tools. For Unix machines, there are many very powerful packages available: AVS, Unigraph, IDL amongst many others are very versatile commercially available packages, whilst excellent public domain packages such as Ace/GR (xmgr & xvgr) and gnuPlot encompass all but the most esoteric plotting facilities.

Details of the techniques of plotting will be covered later in Chapter 7.

Handling errors

It is important in any experimental situation to be acutely aware of sources of error. Errors can be introduced at many stages including statistical random fluctuations in the data, errors and inaccuracies in the measurements, inaccuracies in the calculations (*e.g.* rounding errors) and inaccuracies in other people's results.

Many excellent volumes have been written on error handling, and it is not the intention to go into detail here except to say that a good understanding of errors, their propagation and their minimisation goes a long way to producing reliable results that others will have confidence in.

The handling of instrumental "errors" such as noise is treated excellently in Wayne's book and a classic treatise on the treatment of errors in calculations can be found in Bevington.

5.3 LIMS

Laboratory information management systems (LIMS) are programs, or suites of programs, for the efficient acquisition, storage and manipulation of data. LIMS packages come in all shapes and sizes: some are just specialised databases allowing data to be organised, others have data acquisition programs integrated into them to allow raw data to be retrieved directly from an instrument into the system, but they can all be used to generate reports and create graphical output of the data.

In general, LIMS systems are most often used in analytical laboratories where the same operation is performed on many different samples and where it is necessary to keep track of many thousands of items of data.

6   Computational chemistry

Computational chemistry is using computers to calculate 'chemistry'. The chemistry may be molecular properties, or it might be reaction rates, but it always uses either fundamental properties of a system or a theoretical model as a basis for calculating other properties.

Some would probably say that computational chemistry uses computers in a way in which they were intended – for doing scientific calculations. Others would probably say that it isn't chemistry at all! Nevertheless it has become an important branch of chemistry, especially as the power of computers increases and so the complex calculations required become faster.

There is, of course, no point in delving deeply into the vast field of computational chemistry here. Indeed there is a primer devoted to just this subject. Instead, some of the packages and techniques used in writing programs will be introduced.

6.1 Packages

It was recognised quite a while ago that not everyone wants to write their own programs. Consequently program packages were developed that allowed non-programmers to perform complex calculations. These packages really come in two different forms: those designed for one particular application, and those that are more general in nature. In general though the packages are designed to be easy to use and either operate entirely through a graphical front end, or produce graphical output.

Application specific packages

There are application specific packages for just about any area of chemistry and a list of them would go on for several pages. Most of the packages are derived from programs developed as part of a research project, and while most are in the public domain and freely available (to other academic institutions), a large portion have been further developed as commercial packages. These commercial packages can cost substantial amounts of money: this cost reflects the amount of time spent developing them, to say nothing of the niche market they are intended for.

The platforms on which these packages run vary a great deal. Some are most certainly designed for use on large machines, whilst others perform quite happily on all manner of different architectures. Many of the packages also come as source code, making their use on different machines much easier.

A problem with finding an example to give here is that all the packages are very, very different. They are different not only in what they calculate, but also in the way the programs are controlled and how the results are presented. To some extent this difference is a result of the individual programmers choosing what they think is the best way to approach a problem, but it is also because of the traditions and conventions in a particular field. When using a new package, or even when using a new version of an old package, it is of paramount importance that you read, and understand, the documentation. Some of the non-commercial packages may come with less documentation than is hoped, but nevertheless, read what is available. As a second line of attack on how to use a package, it is always helpful to examine the examples or tests that will inevitably form part of the program – they should not only be run, but they should be altered so that you are sure that you know, and understand, what the program is doing.

Notwithstanding the problems of finding an example, it is important that an example be given! The package chosen is Gaussian. This is a commercial package, but is ubiquitous in physical and theoretical chemistry departments. Gaussian is used to calculate molecular parameters from a theoretical basis. The package runs on many different architectures, ranging from PCs to supercomputers. The actual details of the calculations are not of importance here, just the way that the package is driven.

Gaussian is 'driven' by a data file. This data file contains commands to tell it what calculations to perform along with the data on which to perform that calculation. A typical input file is shown in Fig. 6.1. This file instructs Gaussian to perform an energy calculation on water. The first line instructs the program to use terse output ('#T' – note that the '#' symbol is a command, not a comment!) and that a Restricted Hartree-Fock ('RHF') calculation is to be performed using the 6-31G(d) basis set. The next line is a description of the calculation, and is otherwise unused by the program. The final lines provide a description of the molecule starting with the charge and spin multiplicity (in the case of water it is neutral and a singlet) followed by the element type and Cartesian co-ordinates of each of the atoms in the molecule.

```
#T RHF/6-31G(d)

Water single point energy

0 1
O -0.464 0.177 0.0
H -0.464 1.137 0.0
H 0.441 -0.143 0.0
```

Fig. 6.1 A Gaussian input file to calculate the single point energy of water

Once Gaussian is run with this input file (the exact details on how to do this is site-dependent), a log file is produced. The whole log file is some 175 lines long, so won't be reproduced here, but towards the end of the file is the line:

```
Normal termination of Gaussian 94
```

which indicates that the calculation has completed successfully, and somewhere in the middle of the file is the result:

```
SCF Done: E(RHF) = -76.0098706218 A.U. after 6 cycles
```

indicating that the energy of the system, computed at the Hartree-Fock level, is about −76 hartrees.

General purpose packages

There are many fewer general-purpose packages around than there are application specific ones. These packages usually take input in the form of equations or other recognisable mathematical notation, and perform their calculations on this input. Again, it is impossible to give a 'typical' example of such a package, since they are also all very different, but two of the most common ones are Mathematica and Mathcad. Mathematica claims to be a 'fully integrated environment for technical computing', whilst Mathcad is sold as a 'freeform spreadsheet'. Both these packages, and other similar ones, contain the basic elements for performing numerical, algebraic, symbolic and graphical operations.

Most of these packages run on many different types of machines, although the source code is very rarely made available and so it is necessary to rely on the publisher to make other architectures available. The packages are also usually very graphically orientated – they rely heavily on users being able to input complex equations, and on reporting the results *via* visualisation techniques.

One of the big advantages of many of these packages is that they are capable of performing *symbolic arithmetic*. In other words, they can manipulate equations that involve symbols and not just pure numbers. For instance, it is possible to instruct a package to perform an integration or factor a polynomial. Such capabilities are an immense help when trying to understand a complex set of equations.

As an example of the capabilities of such systems, several Mathematica examples are shown in Fig. 6.2. In these examples we can see that not only can Mathematica be used to perform calculations, but can also perform symbolic arithmetic.

$In[1] := Solve[x^2 + x == a, x]$

$$Out[1] = \left\{ \left\{ x \to \frac{1}{2}\left(-1 - \sqrt{1 + 4a}\right) \right\}, \left\{ x \to \frac{1}{2}\left(-1 + \sqrt{1 + 4a}\right) \right\} \right\}$$

$In[2] := \int (a - b)^\wedge x \, (x - 1)^\wedge a \, dx$

$$Out[2] = -(a - b)(-1 + x)^{1+a} \, Gamma[1 + a, -(-2 + x)Log[a - b]](-(-1 + x)Log[a - b])^{-1-a}$$

$In[3] := ParametricPlot3D[\{u \, Cos[u] \, (4 + Cos[v + u]), u \, Sin[u] \, (4 + Cos[v + u]), u \, Sin[v + u]\},$
$\{u, 0, 4 \, Pi\}, \{v, 0, 2 \, Pi\}, PlotPoints - > \{60, 12\}]$

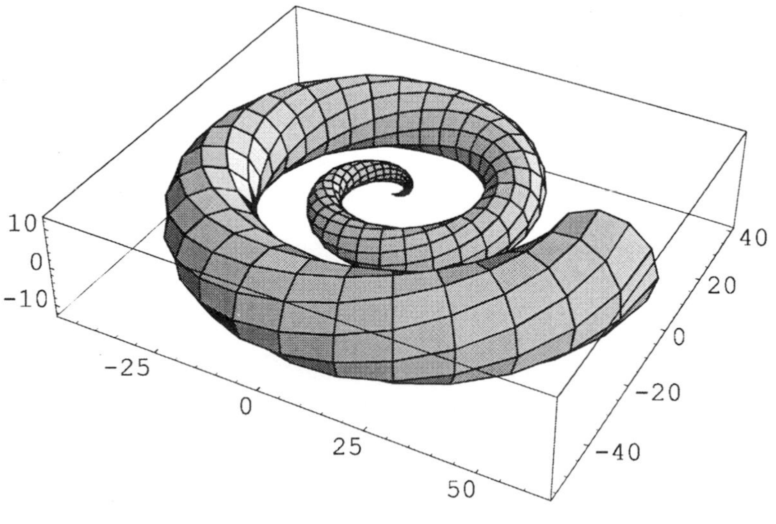

$Out[3] = - Graphics3D -$

Fig. 6.2 Some examples of Mathematica calculations showing both symbolic arithmetic and graphical capabilities.

6.2 Subroutine packages

Although the packages described above can be coerced into performing almost any calculation, they are sometimes slower than purpose written programs. The lack of speed is not necessarily a fault, but more a symptom of using a system that is optimised for general use, and not for your particular application. Sometimes the packages are just not capable of performing the calculations you want. In the end then, in order to perform the calculation, it is necessary to do some programming.

It is possible, of course, to write the whole program yourself, and indeed this will be essential if a new computational system is being developed. However, for more general programming, there are available packages of pre-written subroutines that perform common tasks. These tasks can range from simple matrix operations, to performing numerical integrations or Fourier transforms. There are both commercial and public domain packages, but amongst the best known are ones such as NAG or BLAS. Both of these provide similar functionality, although BLAS is aimed more at basic functions, whereas many of the NAG routines perform very complex calculations. The routines in both these packages are available in both C and Fortran, and so can be called from almost any program. An example of a Fortran program using a NAG subroutine can be seen in Fig. 6.3. Here a subroutine ('f02abf') is used to calculate the eigenvalues and eigenvectors of an N×N matrix. The program is very simple, it merely reads an array, calls the NAG subroutine, and then outputs the results.

The parameters to the NAG routine call provide the method of passing data to the routine (as array A whose size is IA and the order of the matrix N), retrieving the results from it (with the eigenvalues and eigenvectors in arrays R and V) and providing the routine with any workspace it needs (array E). The variable IFAIL allows the routine to pass back to the calling program information on the success of the calculation.

The output of the program, along with the input file, is shown in Fig. 6.4.

```
      INTEGER           NMAX, IA, IV
      PARAMETER         (NMAX=8,IA=NMAX,IV=NMAX)
      INTEGER           I, IFAIL, J, N
      DOUBLE PRECISION A(IA,NMAX), E(NMAX), R(NMAX), V(IV,NMAX)
      EXTERNAL          F02ABF
      WRITE (*,*) 'F02ABF Example Program Results'
C
C Read number of rows/columns
C
      READ (*,*) N
      WRITE (*,*)
      IF (N.LT.1 .OR. N.GT.NMAX) THEN
         WRITE (*,99999) 'N is out of range: N = ', N
         STOP
      END IF
C
C Read array contents into A(,)
C
      READ (*,*) ((A(I,J),J=1,N),I=1,N)
      IFAIL = 1
C
      CALL F02ABF(A,IA,N,R,V,IV,E,IFAIL)
C
      IF (IFAIL.NE.0) THEN
```

Fig. 6.3 Example of program showing how a NAG routine is called

```
              WRITE (*,99999) 'Error in F02ABF. IFAIL =', IFAIL
          ELSE
              WRITE (*,*) 'Eigenvalues'
              WRITE (*,99998) (R(I),I=1,N)
              WRITE (*,*)
              WRITE (*,*) 'Eigenvectors'
              DO 20 I = 1, N
                  WRITE (*,99998) (V(I,J),J=1,N)
   20         CONTINUE
          END IF
          STOP
C
99999 FORMAT (1X,A,I5)
99998 FORMAT (1X,8F9.4)
          END
```

Fig. 6.4 Input and output of the program in Fig. 6.3

Input File:
```
    4
  0.5   0.0   2.3  -2.6
  0.0   0.5  -1.4  -0.7
  2.3  -1.4   0.5   0.0
 -2.6  -0.7   0.0   0.5
```

Output File:
```
F02ABF Example Program Results

Eigenvalues
   -3.0000   -1.0000    2.0000    4.0000

Eigenvectors
    0.7000    0.1000    0.1000   -0.7000
   -0.1000    0.7000    0.7000    0.1000
   -0.5000    0.5000   -0.5000   -0.5000
    0.5000    0.5000   -0.5000    0.5000
```

Another useful source of subroutines is the many books written on mathematical techniques. Often these books contain example code (written usually in Fortran or C) and may contain a diskette or CD. Perhaps the most outstanding example of these is the *Numerical Recipes* series. These books present many numerical techniques and are available in a number of computer languages. I can not recommend these books too highly, and I have yet to see a serious scientific programmer without at least one on their bookshelf.

Some programmers might consider it 'cheating' using pre-written subroutines such as described here. However, it must be remembered that as chemists, our task is not to show our prowess at programming, but to perform a given task. Under these circumstances it is wise to spend time writing code that is peculiar to that task, rather than spend time writing subroutines to perform standard functions.

6.3 Programming techniques

Scientific programming is still performed, to a large extent, in Fortran. This is not necessarily a bad thing. Fortran was designed as a language for scientists and its strengths have stood the test of time. More recently, especially as more work is performed on Unix systems, C is becoming much more in evidence. The availability of subroutine libraries such as NAG and

BLAS in C has done much to help in the acceptance of it into the scientific community. Still more recently, 'modern' languages such as Fortran90 and C++ have been making an appearance.

In the end, it doesn't particularly matter what language a program is written in, so long as it performs the task correctly. The primary factor in deciding which language to use should be the capabilities of the programmer and the facilities available. It is perfectly acceptable to program in Basic on a PC, if the requirements of the project are satisfied.

Good practice

There are a number of general techniques that apply to all languages that it is wise to be aware of. First, good, logical structure to your programs should be maintained. In the early days if programming, when programmers didn't actually interact with the computer and it was necessary to submit a deck of cards or a paper tape to the computer centre for processing, it was a big advantage to get the program correct first time round. To this end, flow charts were used to elucidate the structure of the program before any code was written. These days, when people interact directly with the computer and a program is developed 'on-line', the use of flow charts has fallen into disuse. Nevertheless, it is still good technique to have written down what the structure of your program will be. This will allow you to divide your program into sub-units that will be easier to code. More often than not these sub-units will be individual subroutines, although if the structure dictates, they need not be.

The second thing that should be remembered is to provide comments in the code. Most, if not all, programmers find writing comments a tedious task – they disrupt their flow of thoughts when writing the code, and when they do appear they are likely to be cryptic in the extreme. Nevertheless, it is a task that must be done! Most non-trivial programs will be used on more than one occasion, and they will be modified and developed over time. Comments are invaluable in this development process: they should indicate what each section of code does, and ideally should detail what all the variables are used for. The comments become even more valuable if somebody other than the programmer modifies the code – each subsequent programmer should maintain the comments and indicate what modifications have been performed.

The code should also be presented in a readable manner. This means using variable names that mean something – there are very few programming languages that penalise long variable names, so there is no reason not to use a name such as 'wavelength' rather than 'w'. The program should be laid out in a manner that reflects the structure of the program. For example, the body of loops, or if statements, should be indented so that it is obvious where they begin and end; subroutines should be used; variable declarations should be grouped together; and so on.

Examples of good and bad programming practice are shown in Fig. 6.5. This program uses a Numerical Recipes subroutine (gaussj) to calculate the inverse of a matrix using Gauss-Jordan elimination. It is obvious how much more readable, and understandable, the well-written example is.

Fig. 6.5 Examples of good (bottom) and bad (top) programs. Both these programs perform exactly the same function.

```c
#include <stdio.h>
#include "nr.h"
#include "nrutil.h"

main() {
int j,k,l,m,n; float **a,**b;
srand(1234); n=100;
a=matrix(1,n,1,n);
b=matrix(1,n,1,n);
for (k=1;k<=n;k++)
for (l=1;l<=n;l++)
{ a[k][l]=rand(); b[k][l]=rand(); };
gaussj(a,n,b,n);
for (k=1;k<=5;k++) {
for(l=1;l<=5;l++) { printf("%7.3g ",a[k][l]); }; printf("\n");
};
free_matrix(a,1,n,1,n); free_matrix(b,1,n,1,n);
}
```

```c
/* test program for matrix inversion */
#include <stdio.h>
#include "nr.h"
#include "nrutil.h"

main()
{
        int j,k,l,m,size;
        float **a,**b;

        srand(1234); /* Initialise random number generator */
        size=100;
        printf("Size of array = %i ",size);
/* Create matrices */
        printf("Creating matrices\n");
        a=matrix(1,size,1,size);
        b=matrix(1,size,1,size);
/* Fill matrices with random numbers */
        printf("Filling Matrices\n");
        for (k=1; k<=size; k++)
                for (l=1; l<=size; l++)
                        {
                        a[k][l]=rand();
                        b[k][l]=rand();
                        };
/* invert matrix */
        printf("Starting inversion\n");
        gaussj(a,size,b,size);
        printf("done\n");
/* Print out top 5x5 matrix */
        for (k=1; k<=5; k++)
                {
                for(l=1; l<=5; l++)
                        {
                        printf("%7.3g ",a[k][l]);
                        };
                printf("\n");
                };
/* Free up memory used by matrices */
        free_matrix(a,1,size,1,size);
        free_matrix(b,1,size,1,size);
}
```

Optimisations

Many programs will be fairly trivial and not take up many resources on the machine you are using. However, it may well be the case that the program you have written will take a substantial time to run or will use a lot of memory. If the program will be run often, then it may benefit from some form of optimisation.

Most modern compilers have the ability to perform some optimisations themselves. The optimisation is usually enabled by specifying a command line option in Unix based systems, or in an options menu on Windows-type systems. Different levels of optimisations can usually also be selected – different degrees of optimisation are available as it is usually a trade-off between execution time and compilation time. The sorts of optimisations performed by the compilers are things like taking some calculations out of loops (where the variables involved are not changed during the course of the loop), unrolling loops (*i.e.* repeating code instead of using a loop), moving variables from memory into internal CPU registers when they are accessed often, re-ordering code to make it more efficient, and so on. A few simple examples of this are shown in Fig. 6.6.

```
for (i=0; i<=100; i++)          for (i=0; i<=100; i+=2)
   for (j=0; j<=3; j++)            {
      {                           i+=a[i][0]+a[i][1]+a[i][2]+a[i][3];
      i+= a[i][j]*x;              i+=a[i+1][0]+a[i+1][1]
      };                                   +a[i+1][2]+a[i+1][3];
                                  }
                                i=i*x;
```

Fig. 6.6 Code segments showing some simple optimisations. To the left is the original code, and to the right is some optimisations that the compiler may perform. The inner loop is unrolled completely and the outer loop partially, and the multiplication inside the loop is taken to outside the loops to its logically equivalent place.

Some of these things may seem rather drastic, and indeed they are, so some care must be taken when turning optimisations on. It is always wise to compile and run any program without optimisation first, then add progressive levels and compare the output with the non-optimised version to make sure that the optimisations have not affected the logic of your code. Usually the optimisation procedures used in the compilers are very good, and they do not often cause problems, but it is not unheard of, so some care should be taken.

Although the built-in optimisations are usually very good, they should not 'over-rule' the programmer: if you program poorly, no level of compiler optimisation will get around it. Consequently, you should be aware of good programming in order to get the best out of your program. This is where knowledge of the architecture of the computer pays off. For instance, if you know that you have a 128 kB processor cache on the machine, accessing two memory locations further apart than that will cause un-cached data to be retrieved, thus slowing down the program. If you often access these two items of data together, then if you can arrange for them to be closer together in memory (*i.e.* declare them next to each other) then the overall efficiency of your program will increase.

Multi-dimensional arrays are another area that causes large inefficiencies. Memory is effectively a one-dimensional structure, so multi-dimensional arrays are usually stored in sequential memory locations as shown in Fig. 6.7. The storage may be either row-wise or column-wise, depending on the particular compiler. If, when you come to access the array, the fastest incrementing dimension is not the one that is stored sequentially and the array is large, then large inefficiencies can be introduced.

Fig. 6.7 Row-wise and column-wise storage of an array along with an example of how a 4×4 array would be stored for each case

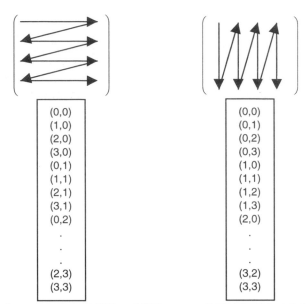

Take for example a 1000 × 1000 array of 32-bit (4-byte) integers on a machine that has a 16 kB cache with 1 kB being read into the cache at a time. Each dimension of the array is about 4 kB long (and the total array is about 4 MB). If the array is stored row-wise, and is accessed in the same way, then each time a 1 kB boundary is crossed, a new chunk of memory will be read in: 4 memory reads will be required for each row and a total of 4 rows can be stored in the cache at any time. If, however, the array is accessed column-wise, and since the 'consecutive' memory locations are then 4 kB apart, each access will cause a memory chunk to be read into cache. After 16 locations have been read, the cache is full, so the next read will replace the oldest data in the cache. Consequently, there will be a memory fetch *every time* the array is accessed. So making sure that the array access is in the correct order will reduce the memory reads from 1 000 000 to 4000. Obviously this size of cache is quite small these days, but the problem occurs as soon as the whole array will not fit in the cache.

As an illustration of the type of problems that may occur, Fig. 6.8 shows two C programs. Each of the two programs performs identical tasks, with only the array access order changed. The version of the program shown in Fig. 6.8a took 16 s of CPU time on an SG Origin2000, whereas that shown in Fig. 6.8b took only 5 s. This difference is also reflected in some of the statistics collected by the system as shown in Table 6.1. Here we see that the number of cache misses is considerably larger for program 'a' compared to program 'b'.

Fig. 6.8 Programs used to illustrate the significance of array access order

```
/* Program 6.8a */
int i,j;
double  a[5000][5000];

void main()
{
for (i=1 ; i<=4999; i++)
 {
  for (j=1 ; j<=4999; j++)
    a[j][i] = i*j;
 };
}
```

```
/* Program 6.8b */
int i,j;
double  a[5000][5000];

void main()
{
for (i=1 ; i<=4999; i++)
 {
  for (j=1 ; j<=4999; j++)
    a[i][j] = i*j;
 };
}
```

Other problems occur with programs that take up large amounts of memory. Most operating systems now use some form of virtual memory. This is memory that is actually disk storage, but appears to be part of the main memory. In this way expensive RAM can be augmented with much less expensive disk. It also means that programs can be bigger than the total memory of the machine. Virtual memory is also used in multi-user machines to free-up the memory used by quiescent programs for use by active programs: this is termed 'swapping out' (since inactive programs are swapped for active ones), and the virtual memory is usually called *swap space*. This is all very useful if you want to create and use programs larger than the real memory. But, the same caveats apply as with cache: if you are accessing two items that are both too large to fit into real memory, then the processor is going to spend most of its time swapping data in and out. These disk accesses are very expensive in terms of time and in severe cases the machine can spend much more time swapping, than it does doing real processing (when it gets into such as state, it is usually called *thrashing* – if you ever see the disk activity of a machine in such a state, you will know why!).

Many systems provide tools to monitor system parameters that affect the efficiency of your program. These parameters are things such as cache hits/misses (*i.e.* how well your program and data fit in cache), swaps (*i.e.* how your program fits in memory), system calls and so on. Careful use of these tools can help in increasing the efficiency of large programs: this will not only speed up your program, but it will also help conserve system resources.

An example of such program statistics is shown in Table 6.1. These statistics are taken from the execution of the programs shown in Fig. 6.8 on an SG Origin2000. Many of the statistics will be fairly meaningless if you are not intimately familiar with the type of machine, but it is obvious, as pointed out above, that there is a big difference in some of the numbers, especially those involving the cache, indicating that there are some inefficiencies in program 'a'.

Table 6.1 Statistics collected from programs in Fig. 6.8 on an SG Origin2000

Event	Program A	Program B
Cycles	1910899872	1001333472
Issued instructions	1059910784	829866256
Issued loads	501103984	350376464
Issued stores	97795392	49987616
Issued store conditionals	0	0
Failed store conditionals	0	0
Decoded branches	50948128	25102320
Quadwords written back from secondary cache	26357808	11887920
Correctable secondary cache data array ECC errors	0	0
Primary instruction cache misses	38288	37968
Secondary instruction cache misses	944	0
Instruction misprediction from secondary cache way prediction table	448	608
External interventions	17504	9040
External invalidations	47808	20016
Virtual coherency conditions	0	0
Graduated instructions	912764032	872576608
Cycles	1910899872	1001333472
Graduated instructions	894354256	871503248
Graduated loads	251947712	278033984
Graduated stores	50164016	49946032
Graduated store conditionals	0	0
Graduated floating point instructions	24808320	25087744
Quadwords written back from primary data cache	34353360	12092224
TLB misses	25951712	6352
Mispredicted branches	6160	10368
Primary data cache misses	24879968	6366928
Secondary data cache misses	3885888	1551776
Data misprediction from secondary cache way prediction table	14613840	336
External intervention hits in secondary cache	15936	7776
External invalidation hits in secondary cache	10928	5184
Store/prefetch exclusive to clean block in secondary cache	80	80
Store/prefetch exclusive to shared block in secondary cache	320	16

7 Presenting information

For all scientists, from school children undertaking classroom experiments to distinguished professors heading multinational research projects, it is necessary to be able to present results in a logical understandable manner. This chapter will attempt to present some of the techniques available using computers to help achieve this aim. Again, it must be stressed that it can not be a comprehensive review of all the programs available, there is simply not enough room here, nor would it be practical considering how quickly such information goes out of date; rather it is a guide to good practice and techniques.

Through necessity, a few packages that the author himself uses will be concentrated on; this is in no means a recommendation that the reader should adopt these packages themselves. The most important thing when using commercial packages is that they should be able to do all the tasks necessary and that the user is comfortable using them. There is little point in using the most recent, all-singing all-dancing word processor, if it takes twice as long to do everything; similarly, there is no point in using a simple, fast, text editor if it is incapable of inserting subscripts or superscripts!

7.1 The armoury

The packages that it is necessary to have available depend, obviously, to a large extent on what sort of work is being presented. It is unlikely that an organic chemist would be able to survive without some form of molecule drawing package; however a physical chemist would find very little use for one. The programs will also depend tremendously on the platform(s) available and the "tradition" amongst those in the same field.

The bare minimum needed is some form of word processor. Indeed, it is surprising what can be done with a modern word processor – they usually include some form of rudimentary drawing package along with some spreadsheet capabilities, and they are virtually all capable of formatting equations. Other useful packages include spreadsheets, drawing packages, graphing programs and, possibly, image manipulation.

Word processors

The ubiquitous word processor, especially on PCs and Macs, is Microsoft Word, although Corel WordPerfect is also very popular. The latter also has the advantage that it is available on some Unix platforms. In reality though, there is little to chose between these two in their capabilities, and the choice will largely depend upon individual tastes (to say nothing of local availability). Although Word and WordPerfect account for the majority of word processing packages installed world-wide, in the scientific community their dominance is rivalled by the TeX document processing package. TeX, or its more popular derivative LaTeX, is particularly favoured amongst mathematicians and those whose chief work is mathematical in nature, such as theoretical chemists.

The chief difference between Word/WordPerfect and LaTeX is that the former are WYSIWIG, whereas the latter is most definitely not. For those who have never come across LaTeX, its style of word processing can seem very strange: the text that is written can be likened to a program source that is compiled to produce the executable code, but in this case the output is not an executable program, but a file containing the formatted document. The text within the LaTeX source file contains instructions to the document processor about how the end document should look. The great advantages of LaTeX is that it is very customisable and the source text can be processed on almost any system since there are LaTeX "compilers" available, free of charge, for virtually every platform. An example of a LaTeX source file and the output it produces is shown in Fig. 7.1.

WYSIWYG – What You See Is What You Get – a term that describes word processors in which what appears on the screen is a direct representation of what will appear in the final printed document (or at least that is the idea – some come closer than others!).

Fig. 7.1 An example LaTeX file, along with the output it produces.

```
\documentstyle[10pt, fleqn]{article}
\newcommand{\ou}{$\ddot{\mbox{o}}$}
\textwidth10cm
\begin{document}
The Schr\ou dinger equation for a particle of mass m
moving in one dimension is

\begin{equation}
\hat{H} \psi = \frac{\hbar^2}{2m}\frac{d^2\psi}{dx^2}
+ \hat{V} \psi = e \psi
\end{equation}
\end{document}
```

The Schrödinger equation for a particle of mass m moving in one dimension is

$$\hat{H}\psi = \frac{\hbar^2}{2m}\frac{d^2\psi}{dx^2} + \hat{V}.\psi = e\psi$$

The lack of WYSIWYG interface for LaTeX has been addressed in a number of ways. Commercial packages, such as Scientific Word for the PC, are available that act as a *front-end* for LaTeX; they look similar to a 'standard' word processor, but produce LaTeX as their output which can then be processed further. Public domain packages, such as LyX are similar and give the same functionality on Unix workstations. These front-ends provide the bulk of the features that are expected of a word processor, but they are lacking in some of the more obscure functions – although this may not be seen as a drawback by some! In addition, they tend not to be as good as the likes of Word and WordPerfect at interacting with other packages on the computer. Their big advantage, as you would expect, is that they are very good at setting mathematical equations.

Style

One of the most difficult things for any writer is the development of a style. A writing style does not come into being fully formed: it can take a month or so of writing to develop. When producing written work of any significant length, my advice is always to go back and re-examine the first pieces of writing to check that the style is not vastly different from that which has developed. You will also inevitably find that your style will change over the years. It is impossible for anybody to say what is the 'correct' style and it is up to the individual to decide what suits them and what is acceptable in the circumstances. Nevertheless, in scientific writing there are a few guidelines which should be born in mind.

First, make sure that the style is not too 'quirky' and difficult to read: individualism can be taken too far. The whole point of scientific writing, be it writing essays or papers or books, is to impart information to the reader. An entertaining style may seem like a good idea, but its novelty wears off after a while when you are trying to search for information. Secondly, grammar is important. Grammar has developed over many years in order to make the meaning of sentences plain and unambiguous. If the grammar or sentence structure is unintelligible, then it will annoy and confuse your readers. Having said that, it is possible to take some liberties with grammar if the meaning is unambiguous: I'm sure that not all of this primer is written in perfect English, but I hope the poor grammar does not detract from the information contained in it! Nevertheless, it should always be remembered that in scientific writing it is the content, not the form, of the writing that is important.

Other things to be careful of when writing are things such as introducing 'lab shorthand' when writing published articles. This means using abbreviations, shortcuts, and so on that others not familiar with your work may not understand. An example may be referring to "singlet delta", when what you really mean is $O_2(^1\Delta_g)$. You should also be aware of using *slang* words: you don't 'zap' things, you 'photolyse' them and so on. This sort of advice may seem obvious, but it is surprising how often such things are seen.

The next thing to be aware of is consistency. There is nothing more annoying and confusing for a reader than for the author to change how things are referred to. For instance, you should be aware of continually changing between using 'ethene' and 'ethylene' or of using 'ml', 'cc', and 'cm^3' interchangeably. Again this may seem like common sense, but often common sense is not at the forefront of your mind when you are concentrating on what you are writing.

Finally, it is important to ensure that what you write is unambiguous. It may seem blindingly obvious to you what you mean, but that does not mean that it is what you have written! Take the short abstract shown below:

"The solution was centrifuged and the supernatant and solid residue extracted. This was then further treated…"

It is unclear here what the 'this' in the second sentence refers to. The rules of grammar say that it refers to the subject of the previous sentence, but it is likely in this case that the writer is referring to either 'the supernatant' or 'the solid residue' and not 'the solution'. The answer is simply to start the second sentence with 'The supernatant…' and all ambiguity is removed.

Equations

In virtually all branches of chemistry you will come across equations. The equations may be chemical or mathematical, but they inevitably form an integral part of the subject. The way that equations are presented depends on whether they are chemical or mathematical, consequently the two will be treated separately.

Chemical equations

Most chemists know how to layout a chemical equation, or at least most chemists have their own ideas about how chemical equations should be laid out. The principle is simple: 'reactants' 'arrow' 'products'. The difficulty really arises when there are many reactions to display together. To some extent the style of presenting groups of equations is up to the author (although journals may proscribe a certain style), so long as the style is maintained. The differences amount to how the various parts of the equation are aligned: some prefer the equation to be monotonically spaced across the page, usually left justified; others prefer the equations to be placed so that the arrows are vertically aligned; yet others prefer each component of the equation to line-up (although this becomes very difficult when there is a mixture of uni-, bi- and termolecular reactions!). Each of these type of alignment are shown in Fig. 7.2.

Fig. 7.2 Different styles of presenting chemical equations:

(a) monotonically spaced, left justified;

$$H + O_2 \rightarrow HO + O$$
$$OH + H_2 \rightarrow H_2O + H$$
$$H + H \rightarrow H_2$$

(b) monotonically spaced, centered;

$$H + O_2 \rightarrow HO + O$$
$$OH + H_2 \rightarrow H_2O + H$$
$$H + H \rightarrow H_2$$

(c) monotonically spaced, centered on the arrow;

$$H + O_2 \rightarrow HO + O$$
$$OH + H_2 \rightarrow H_2O + H$$
$$H + H \rightarrow H_2$$

(d) individual components aligned

H	+	O_2	$\rightarrow$	HO	+	O
OH	+	H_2	$\rightarrow$	H_2O	+	H
H	+	H	$\rightarrow$	H_2		

In general, any of these formats, and other similar ones, are acceptable, so long as one style is chosen and stuck to: style (a) is generally the easiest to produce of these, whilst style (d) probably looks best (but that is just personal opinion). The advantage of these simple formats is that they can be produced using nothing more complicated than the normal text functions of a word processor.

Further complications arise when more information is needed within the equations itself. For instance, it may be necessary to put something above the arrow:

$$H_2 + \tfrac{1}{2}O_2 \xrightarrow{\text{Pt}} H_2O$$

It is usually not easy to produce such constructs using normal word processor features; however the above example was obtained fairly easily using a mathematical equation editor. Reactions that are more complex or larger reaction schemes may need to be created using a drawing package (see later) and *imported* into the document.

Mathematical equations

Mathematical equations are probably one of the most complex constructs that people will come across during the normal course of document processing. Most modern word processors have some form of equation editing function built in to them, indeed some, such as LaTeX, have mathematical typesetting as their primary function.

The equation function of some of the word processors is actually a separate program, and it is entirely possible to use the equation editor from one program to produce the equations in a different one.

There are two basic methods that programs use to produce equations: either graphically (using icons to pick the elements of the equation) or as a text description of the equation. The Microsoft Equation editor (as used primarily by Microsoft Word) and the later versions of WordPerfect, amongst others, use a graphical method, whereas LaTeX, earlier versions of WordPerfect, and many Unix based document processors use the text method.

Many of the textual description packages have as their predecessor the 'eqn' equation editor that is part of the 'roff' package available on just about any Unix system. In this system, the principle is to describe the equation as you see it. For instance, to produce the equation:

$$x = \left(\frac{a+b}{\alpha + \beta} \right)^2$$

would require the 'program'

```
x = left ( { a + b } over { alpha + beta } right ) sup 2
```

As can be seen, the equations are built up by grouping individual elements together using braces, and in this way, quite complex equations can be constructed. However, some people, especially those who do not have a mathematical background, find this textual descriptive method of producing equations very confusing.

The graphical method of producing equations is usually much simpler to understand: the particular construct or symbol you wish to insert is simply clicked on. Even non-mathematically based people can construct complex equations just on the basis that it 'looks right', without any real knowledge of what it is they are producing! Fig. 7.3 shows the same equation as above being produced in Microsoft Equation 3.0. As can be seen, this screen shot is taken just as the β is being inserted.

Equation numbering

There are two schools of thought when it comes to numbering equations: should the mathematical and chemical equations be numbered together, or separately? There is not really an easy answer to this! It depends on the context in which the document is produced and on any style proscribed. In general most documents will be largely chemical or largely mathematical,

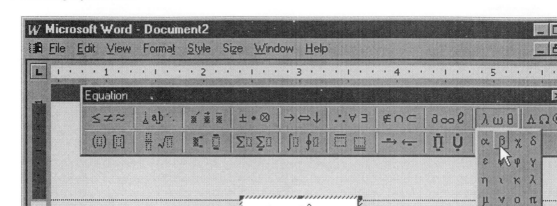

Fig. 7.3 Producing an equation in Microsoft Equation 3.0

and in these cases the minority equation can be labelled with a prefix, often 'E', so that, for example, the maths equations will be (1), (2), (3), … and the chemical equations will be (E1), (E2), (E3), … . Using this type of scheme will remove any ambiguity.

Of course, if the document you are producing is in chapters, such as a thesis or a book, then the equations, and indeed everything else, should be prefixed with the chapter number.

Diagrams

Diagrams are an important part of any scientific text. It is much easier to explain concepts in the form of diagrams than in words and the old adage of "a picture is worth a thousand words" is particularly pertinent in science.

The choices when producing diagrams using a computer are just as wide as for producing text. Many word processors come as part of a suite of programs that include a drawing package, whereas others, such as Word, have some form of drawing capability built-in. However, these packages are usually only limited in their capabilities, and it is often worthwhile using a full package such as CorelDraw or Adobe Illustrator. One drawback with such large and powerful packages is that it takes some while to become proficient with their use, but the time taken practising is usually worthwhile considering the end result.

One thing that must be remembered when drawing scientific diagrams is that you don't need to be an artist. The idea is to present a clear interpretation of the concept you are trying to portray, there is no need for it to be a work of high art. For instance, if it is necessary to draw an apparatus, there is usually no need for it to be an accurate scale drawing, indeed the individual components may not necessarily be in proportion since some parts may need to be emphasised over others, the only thing that is necessary is that it is logically correct. The key really is to try to make the diagrams uncluttered and to some degree self explanatory.

Notwithstanding matters of style, there are a few guidelines that should be adhered to. First, don't make diagrams too small, or try to cram too much detail in. Not only will it make the illustrations more difficult to produce and much more difficult to understand, but also, if the work is to be published, it will make accurate reproduction difficult. On a similar theme, don't be afraid to use quite heavy lines. Thin lines may look nice on the original, but after a couple of photocopies, they will disappear! Secondly, clearly label things; if there is not much room, then use letters or numbers and include a key. The font used for the labels is also quite important: avoid cursive texts if possible, and if the style allows, use a sans-serif font such as Helvetica or Arial. Next, if it is possible to use colour, do so. Colour can significantly enhance the comprehensibility of a diagram and it allows specific items to be highlighted very effectively. But, don't go overboard with colour: garish colours splashed all over the diagram will only distract the reader from the point of the illustration. Finally, keep it simple. If a diagram is to illustrate a particular point or aspect of the work, then make sure that that is all that it does: remove any extraneous detail that is not pertinent to that particular aspect. A sample diagram is shown in Fig. 7.4; this particular diagram is a schematic representation of a photo-oxidation – gas chromatograph system. It is obvious that the diagram is not to scale, and that unimportant details are omitted. For instance, the electrical circuitry controlling the photolysis lamps

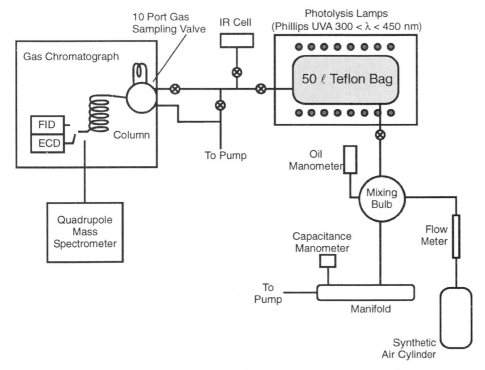

Fig. 7.4 Sample diagram showing some techniques used in making clear illustrations

is very important; however it is irrelevant to this particular diagram, so it is omitted. Similarly, the Teflon bag is not drawn to scale, because although it is of prime importance in the experiment, drawing it much larger does not add any more information to the diagram.

A particular aspect of illustration, which is unique to chemistry, is the depiction of molecules. There are two main types of molecular illustration: structures and molecular modelling. Structural representation is usually employed by organic chemists to show the structure of complex molecules. Special packages are available to aid in producing these diagrams probably the best known examples are programs such as ChemDraw or Isis/Draw, but most general purpose drawing packages can be coerced into producing perfectly acceptable diagrams. The advantage of the specialised packages is that they have the various bonds and ring structures pre-programmed into them, often they are also aware of chemistry and so will help in constructing the diagram. Fig. 7.5 shows a screen shot from the Isis/Draw package while a molecule (adamantane) is being produced. Various predefined structures can clearly be seen on the top menu bar, whilst the side bar shows various tools used in the construction of the molecule.

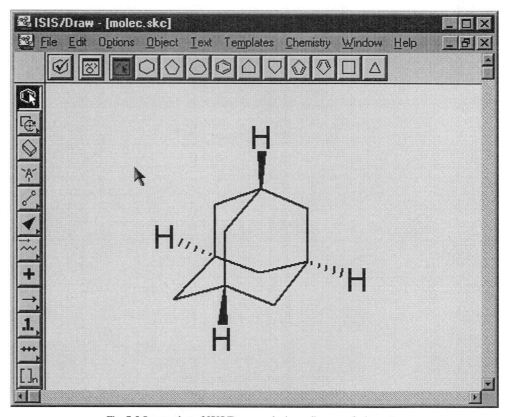

Fig. 7.5 Screen shot of ISIS/Draw producing a diagram of adamantane

Molecular modelling packages are designed to give a realistic three-dimensional picture of a molecule. They may either work as a molecular designer where molecules can be built up by hand or be given the co-ordinates of the individual atoms and so produce a picture. There are usually different ways displaying each molecule: such as stick, ball-and-stick or space-filling, with different three-dimensional effects applied to them. Some examples of different representations of 2-methyl butadiene are shown in Fig. 7.6. Of course, it is impossible to give a good representation of these inherently 3D images on a page; the only way to get a good feel for a molecule in three dimensions is to use a computer and manipulate the image directly, preferably using stereoscopic imaging.

Fig. 7.6 Examples of different representations of 2-methyl butadiene. These pictures were produced using Cerius² from MSI Inc.

a) Stick model showing the bond multiplicity, but no 3D effects.

b) Cylinder model. This particular representation is useful for large molecules such as DNA as it gives fairly good 3D representation and is capable of distinguishing between different atoms.

c) Ball and stick model. This is one of the classic representations of a molecule: the size of the balls gives an indication of the relative size of each atom.

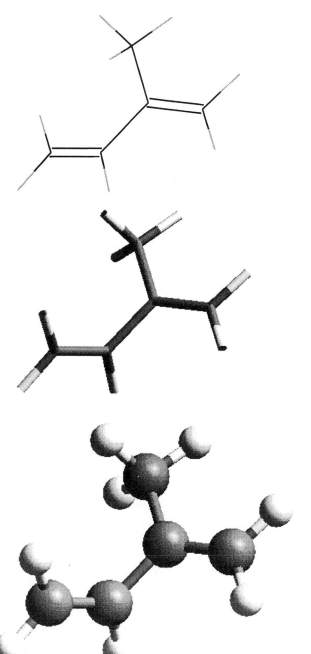

d) Ball or space-filling model. This is the other classic representation of a molecule. The principle is that each atom takes up approximately the correct relative amount of space in the molecule.

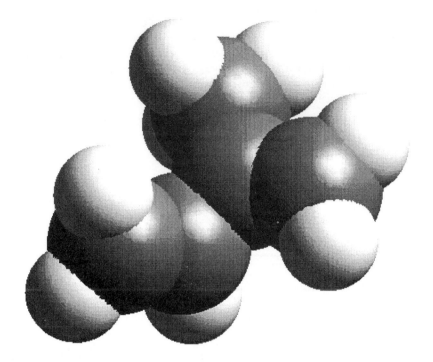

Tables

Tables are a convenient way of presenting a large amount of information in a small space. In scientific writing tables usually consist of lists of numbers whereas elsewhere they often consist of text, and as such their construction needs some attention. Most word processors have many built-in functions to help in producing tables, but as usual it is the author's control of these functions that is critical, not just their presence.

The function of a table is to impart information, so care must be taken that that information is presented unambiguously. Of particular importance in scientific tables are the column headings: these must accurately label what is in the column. Often the numbers in the column will have units and rather than repeat the units for each individual entry, the units are often put in the column heading; similarly if all the values are of a similar order of magnitude, a multiplying factor may be put in the column heading. The style of labelling the column may be proscribed by the ultimate destination of the written work, but where it isn't, then it is up to the author how to present the information. There are two main styles in use: either the multiplier and units in parenthesis after the label, or a *quantity calculus* style is used.

In general, the elements of a table are not presented in their own individual boxes, despite what the word processors would have us believe. Most often horizontal rules only are used, and then only to separate the headers from the main body of the table and to delimit the top and bottom of the table as a whole. Sometimes horizontal rules can be used to separate sections of a table, but that is often more effectively achieved using a blank line. Vertical rules can be used to separate columns, but again, it is often better to omit these lines and merely ensure that there is sufficient space between the columns to make the division obvious.

If possible, a table should not be split across a page boundary. Ensuring such a thing does not occur is usually quite easy with modern word processors, although the method of automatically ensuring such a thing varies from program to program. If it is inevitable that a table is split across a page, for example the table may be longer than a page, then it is important that a single cell is not split across the page, and that the column headings are repeated at the top of the second page. Both Word and WordPerfect have facilities for specifying which row is the header row and automatically repeating this at the top of every page.

As with all other aspects of writing, once a style of table has been decided on, it should be maintained throughout the whole document. Changing styles of such prominent items as tables only makes the work look shoddy and will taint the view of the reader when it comes to judging the content.

Fig. 7.7 below shows good and bad examples of tables. It is obvious how much clearer and uncluttered the second table is. Note also the use of scaling factors for the numbers in right hand columns, the trend is much more apparent than when non-scaled numbers are used.

Wavelength	$\sigma(\lambda)_{ClClO2}$	$\sigma(\lambda)_{Cl2O4}$	$\sigma(\lambda)_{Cl2O6}$
260	5.30×10^{-18}	3.1×10^{-19}	1.268×10^{-17}
265	5.50×10^{-18}	2.2×10^{-19}	1.374×10^{-17}
270	6.60×10^{-18}	1.4×10^{-18}	1.439×10^{-17}
275	8.30×10^{-18}	8.8×10^{-20}	1.464×10^{-17}
280	1.05×10^{-17}	5.5×10^{-20}	1.457×10^{-17}
285	1.26×10^{-17}	4.0×10^{-20}	1.410×10^{-17}
290	1.43×10^{-17}	2.7×10^{-20}	1.316×10^{-17}
295	1.50×10^{-17}	2.2×10^{-20}	1.169×10^{-17}
300	1.47×10^{-17}	1.7×10^{-20}	1.011×10^{-17}
305	1.33×10^{-17}	1.2×10^{-20}	8.550×10^{-18}
310	1.12×10^{-17}	7.0×10^{-21}	7.080×10^{-18}
315	8.70×10^{-18}		5.830×10^{-18}
320	6.30×10^{-18}		4.770×10^{-18}

Fig. 7.7 Example tables. Compare the two examples: they both provide the same information; however the layout of the lower one is much clearer and the trends in the values are much more obvious.

Wavelength	$\sigma(\lambda)$ (10^{-20} cm^2 molecule^{-1})		
(nm)	$ClClO_2$	Cl_2O_4	Cl_2O_4
260	530	31	1268
265	550	22	1374
270	660	14	1439
275	830	8.8	1464
280	1050	5.5	1457
285	1260	4.0	1410
290	1430	2.7	1316
295	1500	2.2	1169
300	1470	1.7	1011
305	1330	1.2	855
310	1120	0.7	708
315	870		583
320	630		477

Graphs

Graphing was first introduced in Chapter 5, where some of the programs available for producing graphs were introduced. However, graphs are not just used in the primary analysis of results. They are a useful tool in illustrating written work, and usually they go hand-in-hand with tables; in fact there is no better way of illustrating a trend (or otherwise) in data than displaying it on a graph.

As with any other diagram, care must be taken to ensure that a graph is clear and illustrates the point it is intended to. Again, there are a few guidelines to be aware of when creating a graph. First, make sure that the axis scales are relevant: this does not, necessarily, mean that the graph must fill all the space, but that they are sensible considering the point you intend to make. For instance, if the purpose of the graph is to illustrate that a set of data is invariant, then there is little sense in choosing the scale of the graph such that the points do not look invariant! But, you mustn't use the scale to hide things you would prefer weren't there.

Secondly, there should be a reasonable number of tick marks: an axis that spans from 0 to 100, should not have tick marks every unit, nor should it have tick marks every 100 units. As a general rule there should be more than three, and ten or less tick marks: A 0–2 axis should have major ticks every 0.5 units, with minor ticks every 0.25, whereas a 0–8 axis would have ticks every unit, with minor ticks every 0.5. As always though, this is not a hard and fast rule: if it is relevant to the purpose of the graph, then tick marks can be at whatever spacing you choose. Most major graphing packages will make life easier by making sensible choices for you when the graph is created.

On the subject of axes, it is very important that they are labelled correctly. This doesn't just mean what the axes represent, but also what units are used. As with tables, the tick labels should be scaled so that they lie in a sensible range (usually 0–10), and both this scaling and the units should be indicated on the axes using methods similar to that described for tables.

The symbols and lines on the graphs should also be chosen with care. When more than one data set is being plotted different symbols should be used for each data set. These symbols are usually chosen from one of the following: ■ ● ▲ □ ○ △ , but other symbols can be used if necessary. Although there are different types of lines available, it is not usual to change the line type for each data set – it is usually obvious which line goes with each data set. Where it is useful to use different line types is when there is more than one line associated with each data set. For example, if the purpose of the graph is to show how different parameters fit a set of data, then it might be sensible to plot the best fit data as a solid line, and the others as different types of dashed line. When different line styles are used, the distinction between them should be obvious – there is no point in using different length dashes when you need a micrometer to tell the difference between them. The size of the data points and thickness of the lines should also be carefully chosen. There is little point in having data points so small that it is impossible to see what the symbol is, but conversely it is unwise to make them so large that they obscure the purpose of the plot. The lines should be thick enough to be plainly visible but without intruding and detracting from the data points and without breaking up when the figure is reproduced. Finally, the lines should appear to be below the data points.

If it is possible, then colours can be used with good effect in graphs. But again, it is unwise to "go overboard" about it: rainbow-hued graphs would probably detract from the intended message rather than enhance and highlight a particular aspect of the data. However, if the graph is to be published, then it is unlikely that colours can be used in something like a graph unless it was of absolute necessity.

Fig. 7.8 shows "good" and "bad" graphs. These figures illustrate some of the points alluded to above. Notice in the first graph how the symbols used for the two sets of points are basically indistinguishable, and that it is virtually impossible to see the trends where the two sets overlap.

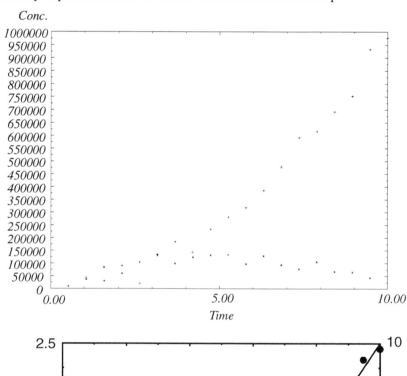

Fig. 7.8 Examples of good (lower) and bad (upper) graphs. The "good" graph is considerably clearer.

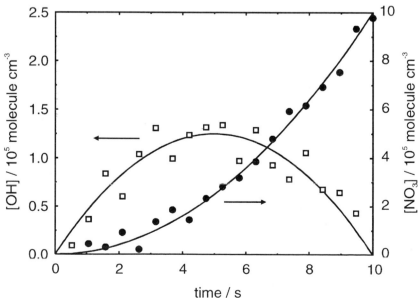

A further useful technique is shown in Fig. 7.8. When sets of data are on clearly different scales, then two different axis scales can be used. The scale that corresponds to each data set should be clearly indicated; in this case, arrows are used to point to the scale that should be used. A different way of achieving the same aim is to 'multiply' one data set by a scaling factor, then indicate on the plot what that factor is. Obviously, if the aim of the graph is to highlight the differences in scale, then it is probably best not to use these techniques.

8   Chemistry and the Internet

Chemistry has benefited greatly from the Internet and there are countless sources of chemical information available on-line. Consequently, it might be thought that any book on computers in chemistry would have to contain a comprehensive section on chemistry on the Internet. However, the evolving nature of the Internet creates a big problem: as soon as such a section is written, never mind published, it is out of date. So, I'm afraid that this author has taken the approach that it is better to keep quiet rather than be wrong!

A few hints though are in order. As has already been said, there is a vast amount of information available on the Internet on all aspects of chemistry. A search for the word 'chemistry' on one of the World Wide Web search engines produced over 4 million web pages, and in all those pages is the information you are probably looking for. So, it is worthwhile getting familiar with the advanced search facilities of any source of information you use, it can reduce the number of dead-ends you follow and so increase the likelihood of finding what you need.

A further warning about the World Wide Web is also needed: *there is no review process for publication on the web*. This means that anybody can put any information they want on a web site and you have little idea as to whether it is based on good quality research or is some lunatic's pet theory.

Chemistry is a subject that benefits greatly from the graphical aspects of the World Wide Web: molecules and so on can be displayed in three dimensions and animated in ways not possible with standard paper publications. Indeed, many paper journals now have 'Internet Editions' just so that advantage can be taken of these graphical capabilities and more recently purely Internet journals have been appearing. However, the protocols and formats of the data needed to produce the advanced chemical graphics is evolving, and it is inevitable that a particular computer may not be able to display a particular graphic because it does not have the right software installed. This, unfortunately, is the price that it is necessary to pay for working in an evolving medium! The article containing the graphics should give you hints as to where to find the required viewing software, or it may be necessary to do a search for it.

Finally, if the thought of searching through 4 million web pages is daunting, then a good place to start is a university web site. Virtually all chemistry departments will have their own web site, and that web site will probably contain links to other places – after a bit of 'surfing' you should be able to find the information you want. No doubt, if you look hard enough, you may even find mention of this primer somewhere!

Bibliography

R.P. Wayne. *Chemical instrumentation* (Oxford Chemistry Primers 24). Oxford University Press, Oxford, 1994 – This primer gives an excellent introduction for chemists to analogue electronics and instrumentation in general.

P. Horowitz, and W. Hill. *The art of electronics* (2nd edn). Cambridge University Press, Cambridge, 1989 – A very valuable book on electronics for all levels.

W.H. Press, B.P. Flannery, S.A. Teukolsky, and W.T. Vetterling. *Numerical Recipes. The art of scientific computing.* Cambridge University Press, Cambridge, 1996 – The new (1996) edition of this book is available in C, Fortran 77, and Fortran 90 versions, whereas the first (1988) edition was produced for C, Fortran, Basic, and Pascal. No self-respecting scientific programmer should be without at least one copy on their shelf.

P.R. Bevington. *Data reduction and error analysis for the physical sciences.* McGraw-Hill Book Company, 1969 – A good text on numerical techniques and statistics.

G.H. Grant and W.G. Richards. *Computational chemistry* (Oxford Chemistry Primers 29). Oxford University Press, 1995.

And finally, if any programming in C is undertaken, you must have a copy of:

B.W. Kernighan and D.M. Ritchie. *The C programming language.* Prentice Hall, 1978 and 1988 – The first edition describes 'K & R' C, whilst the second edition describes ANSI C. Both editions are very useful.

Index